TEXAS TASK FORCE 1

Urban Search & Rescue

BUD FORCE

Foreword by G. Kemble Bennett

Texas A&M University Press • College Station

Copyright © 2011
by Bud Force
Manufactured in China
by Everbest Printing Co.,
through Four Colour Print Group
All rights reserved
First edition

This paper meets the requirements
of ANSI/NISO Z39.48<H>1992
(Permanence of Paper).
Binding materials have been
chosen for durability.

LIBRARY OF CONGRESS
CATALOGING-IN-PUBLICATION DATA

Force, Bud, 1979–
  Texas Task Force 1 : urban search and rescue / Bud Force ;
foreword by G. Kemble Bennett.—1st ed.
      p. cm.
  Texas Task Force One
  Includes index.
  ISBN 978-1-60344-288-6 (hc-trade cloth : alk. paper)—
  ISBN 978-1-60344-383-8 (e-book)
  1. Texas Task Force 1. 2. Disaster relief—Texas.
3. Emergency management—Texas. 4. Search and rescue
operations—Texas. 5. Rescue dogs—Texas. I. Title.
II. Title: Texas Task Force One.
HV555.U62T4135   2011
363.34'8109764—dc22
2010049955

Adapt and Overcome
—Urban Search and Rescue motto

Contents

Foreword by G. Kemble Bennett   ix
Acknowledgments   xi

1. History and Overview   1
2. Training   9
3. Equipment   31
4. Responders   52
5. Canines   82
6. Disasters   100

Afterword by J. Robert McKee   139
Index   141

G. Kemble (Kem) Bennett, Texas Task Force 1's founder, stands amid the devastation of the World Trade Center collapse a few days after the September 11, 2001, terrorist attack in New York City.

Foreword

Specific events can be pivotal moments in our lives. For me, one of those moments was the 1995 bombing of the Alfred P. Murrah Federal Building in Oklahoma City. At the time, as director of the Texas Engineering Extension Service, it was my responsibility to verify that emergency responders in the state of Texas were prepared and equipped to respond to disasters. Until that point, terrorist attacks of such magnitude had not happened in the United States. The tragic events in Oklahoma City raised a frightening question in my mind. Were we truly prepared to respond to terrorist attacks?

I immediately began discussions to create a team dedicated to responding to disasters, both natural and human-made. As I worked with the emergency response community and state government, the concept of a properly trained and equipped emergency response team became a reality when Texas Task Force 1 was formed in 1997.

Initially, the task force was assigned to serve as an urban search and rescue resource for the state of Texas. We were focused on our purpose and dedicated to excellence in all our endeavors. Our program quickly became one of the finest in the nation, and within a couple of years Texas Task Force 1 was named one of the twenty-eight elite national response teams under the Department of Homeland Security's Federal Emergency Management Agency.

Since then, our journey has included responding to some of the most significant disasters in American history. From search and recovery at the World Trade Center towers to water rescue in New Orleans after Hurricane Katrina to the search of a massive expanse of the state following the Space Shuttle *Columbia* explosion, Texas Task Force 1 has been at the forefront of search and rescue at these disasters.

But the disasters are only the backdrop of a much more important and poignant message: In the midst of darkness and despair, there are men, women, and trained canines who have made it their life's work to respond when tragedy strikes. They have committed themselves to extensive, rigorous training with dedication and discipline so that there are no others better prepared to succeed when called upon to serve.

It takes a special kind of person to rush into dangerous environments to help others. Emergency responders see human suffering on a massive scale yet must remain focused on the critical tasks at hand. As you read their stories on the pages that follow, I hope you gain a better understanding of what Texas Task Force 1 is and, more important, who the people are who so selflessly serve.

This is the story of America's finest. I hope you enjoy learning about Texas Task Force 1 and join me in my appreciation of all those who have been, are, and will be a part of its tremendous history and legacy.

—G. Kemble Bennett, PhD, PE
Vice Chancellor and Dean of Engineering at Texas A&M University
Founding Director of Texas Task Force 1

Acknowledgments

Many people participated in the creation of this book. I would like to thank Texas Task Force 1 (www.usar.tamu.edu) and the Texas Engineering Extension Service (www.teex.org) for their support and blessing to create a work that shows what they do to assist the state of Texas and the United States of America on a daily basis. I would like to thank God for the ability, my wife for the support, and my mother for the education that has allowed me to complete this work, which at times seemed like a huge research paper of sorts. Thank you to my family for everything you do.

I would also like to thank G. Kemble Bennett for his fortitude in creating an organization as great as Texas Task Force 1. Thank you to Bob McKee, director of Texas Task Force 1, without whose guidance and contribution this book would have never come into existence.

Thank you to Billy Parker, Susann Brown, Chuck Jones, Jeff Saunders, Matthew Minson, and Jim Yeager for your time in providing interviews and extensive information about Texas Task Force 1. Thank you Brett Dixon.

Thank you to the rest of Texas Task Force 1, both present members and those who have passed on, for what you do and have done to serve selflessly for others.

TEXAS TASK FORCE 1

1 : History and Overview

Texas Task Force 1 (TX-TF1) is more than a search and rescue team. It is a living organization whose heart comprises a diverse group of men and women who come from all walks of life and industries. The team is forged from the firefighters whose primary role is saving people from harm or death; from the doctors who will repeatedly risk their own lives to save others; and from the canine handlers and other civilians who will put themselves and their best friends in danger in order to remove others from it. Most important, it is forged from the common line of thinking to always place oneself second and one's fellow human beings first, to always think of those in distress as one's own family members, to always assist selflessly when needed.

On paper, Texas Task Force 1 is responsible for search and rescue under the Texas Division of Emergency Management while simultaneously serving as one of the federal teams under the Department of Homeland Security and the Federal Emergency Management Agency (FEMA). Overall, TX-TF1 is a multidisciplined, elite entity capable of ground, water, and air search and rescue deployments throughout the nation. As of this date, Texas Task Force 1 is

A C-130 military aircraft stands by as Texas Task Force 1 members haul supplies to an evacuation staging area for medical special needs patients two days before Hurricane Gustav's presumed landfall near New Orleans, Louisiana. The task force, in coordination with military air support, was able to relocate hundreds of patients to a specified safety zone.

one of the most-deployed urban search and rescue teams in the United States and was used after both the World Trade Center attacks on September 11, 2001, and the largest natural disaster on record in our country—Hurricane Katrina. The team's more than 450 highly trained members come from approximately sixty organizations throughout the state and include firefighters, canine handlers, medical specialists and doctors, heavy-equipment operators, structural engineers, and people from various other disciplines.

Afternoon sun splashes on task force members as they plan a wide-area search.

> What is search and rescue? "Search and rescue" is a comprehensive term that includes a broad range of capabilities, such as initially evaluating disaster sites and victims' needs and then providing feedback to local, state, and federal officials in a systematic manner. From there, it includes the actual search and rescue operations in venues ranging from collapsed buildings to raging rivers. Once victims are found, a medical team is available to provide field care and transport. Finally, a hazardous-materials component surveys, evaluates, and contains hazardous materials for the safety of the victims and the search and rescue team itself. In its totality, search and rescue is a diverse and creative response tool that can assess and initially address just about any catastrophic situation one may confront.

The initial concept for Texas Task Force 1, founded in 1997, was the idea of G. Kemble Bennett, dean of engineering at Texas A&M University and, at that time, director of the Texas Engineering Extension Service. The 1995 Oklahoma City bombings at the Alfred P. Murrah Federal Building ignited a desire in Bennett to create a team in his home state that would be fully prepared and ready if a similar disaster should strike Texas. He immediately consulted with numerous peers and formed an advisory board to bring expertise from across the country and set the foundation for the new organization. Eventually, his efforts led to the creation of the team as we know it today; Texas Task Force 1 held its first organizational meeting on February 14, 1997.

Three years later, the task force officially joined the roster of twenty-eight federal teams under the National Urban Search and Rescue System. In addition, it became one of the first six of those teams with the capability to respond to attacks by weapons of mass destruction.

Each search and rescue team under the federal

umbrella is sponsored by a respective agency or organization. The sponsoring entities assist in providing funding, members, and training. In many cases, the sponsor is a fire department. The sponsor of TX-TF1 is the Texas Engineering Extension Service, or TEEX. The agency is regarded as a worldwide leader in training for emergency response and teaches around two hundred thousand people from more than sixty countries each year. TEEX, a member of the Texas A&M University System, one of the largest higher-education organizations in the world, is a state agency that runs everything from the world's largest live-fire firefighting school to unexploded ordnance training programs, where students learn to safely disarm explosives.

All statewide requests to activate Texas Task Force 1 begin at the local government level and are communicated through a series of assets to the Texas Division of Emergency Management. If approved, activation orders are immediately sent and the requested team members then rendezvous at the TX-TF1 headquarters. After medical screening and various levels of processing, the task force deploys to the scene of the disaster with enough equipment and supplies to be self-sufficient for up to ten days of operations.

Responders must always remain aware of their surroundings to avoid becoming victims themselves.

Texas Task Force 1 can basically be configured for three types of rescue missions depending on the scale of the disaster.*

Information provided by the Texas Engineering Extension Service.

1. A large-scale search and rescue team capable of responding to attacks by weapons of mass destruction includes these components:

- An entire command and planning staff for twenty-four-hour-day operations or a staff divided between two divisions for coverage of two separate incident sites
- Communication technologies, procedures, and staffing
- Reconnaissance and search capabilities such as visual and audio technologies and canine search teams
- Rescue capabilities for large-scale structural collapses
- Medical capabilities for incident victims and the team itself
- Hazardous-materials capabilities for detection and protection for continued operations in a contaminated environment

2. A reconnaissance/light rescue team includes these components:

- An entire command and planning staff for twelve-hour-day operations at one incident site
- Communication technologies, procedures, and staffing
- Reconnaissance and search capabilities such as visual and audio technologies and canine search teams
- Rescue capabilities for smaller and lighter structural collapses
- Medical capabilities for treatment of search and rescue team members
- Hazardous-materials capabilities only for detection

3. Various combinations of water-rescue teams include these components:

- Command and planning staffs
- Reconnaissance and search capabilities such as visual and audio technologies and canine search teams
- Boat squads composed of five responders per squad
- Medical capabilities for treatment of search and rescue team members
- Hazardous-materials capabilities for detection and sometimes protection for continued operations in a contaminated environment depending on the configuration

Advanced techniques are used throughout search and rescue operations. These swift-water rescue specialists are righting a capsized raft on the Comal River in Central Texas.

One of the most prestigious in the United States, Texas Task Force 1's water-rescue program began in 2000 with the first team in the nation to operate on a statewide level. The strike team members are ready to deploy within one hour of notification and work directly with Texas Military Forces and other state and federal agencies for needed air and ground support. The team maintains a multimillion-dollar equipment cache weighing in excess of fifty tons, with gear including everything from specialized victim location devices and medical triage tools to boats and hydraulic jacks.

Texas Task Force 1 responds to these types of situations:

- Hurricanes
- Tornadoes
- Major flooding
- Wide-area searches
- Weapons of mass destruction terrorist attacks
- Chemical, biological, radiological, and nuclear explosion incidents
- Large event pre-positioning, such as at the Olympics and national political conventions

History of TX-TF1's major deployments:

1998—Del Rio, Texas, flooding
1999—Texas A&M University Aggie bonfire collapse
2001—Tropical Storm Allison/Houston, Texas, flooding
2001—New York City World Trade Center attacks
2003—Space Shuttle Columbia recovery
2004—Hurricane Ivan
2005—Hurricane Katrina
2005—Hurricane Rita
2008—Hurricane Ike
2010—Rio Grande, Texas, flooding

Texas Task Force 1 works with various agencies to rescue victims from a downtown New Orleans hotel after Hurricane Katrina.

Fact Sheet: Texas Task Force 1

CAPABILITIES
- Texas Task Force 1 members respond to mass casualty disasters and are trained and equipped to locate and extricate victims trapped in collapsed structures, confined spaces, and trenches in highly populated areas.
- The task force is capable of responding to state and national disasters, including earthquakes, hurricanes, widespread tornadoes, floods, and technological and terrorist events.
- The team is designed to be logistically self-sufficient for the first seventy-two hours of operation and is able to function for up to ten days.
- Texas Task Force 1 can be further divided into twenty-eight-member teams capable of responding to light structural collapses and general rescue situations, as well as into five-member water-strike teams for flood response.

PERSONNEL
- Texas Task Force 1 is made up of more than 450 emergency response personnel from approximately sixty organizations and departments across the state.
- Task force members are capable of deploying within four hours and are divided into three teams (red, white, and blue) of seventy members each, which are each on a thirty-day rotational call, and an additional development team.
- Each team has five components: a command structure; a rescue group; a medical group; a logistics planning group; and a search group, including canine search teams.

EQUIPMENT
- Texas Task Force 1 maintains a $5 million state-of-the-art equipment cache of more than twelve thousand items weighing in excess of fifty thousand pounds.
- Equipment includes hydraulic jacks; rams; shoring; high-tech listening devices; hazardous-material monitoring equipment; specialized victim location devices; breaching, breaking, and lifting equipment; and medical and triage equipment.
- Swift-water rescue equipment includes boats, personal flotation devices, and advanced communication equipment.

TRAINING
- Texas Task Force 1 trains at Disaster City® in College Station. Disaster City is a fifty-two-acre training facility that has been designed to simulate various levels of disaster and structural collapse. Filled with full-scale collapsible buildings, the facility features a wide array of infrastructure found in most communities, as well as collapsed strip malls and derailed passenger trains.
- The team's swift-water rescue training occurs at high-water-current sites throughout Texas, ranging from water parks to rivers.

HISTORY
- Texas Task Force 1 was formed after the 1995 bombing of the Alfred P. Murrah Federal Building in Oklahoma City, Oklahoma. The task force held its first organizational meeting on February 14, 1997.
- The task force joined FEMA's National Urban Search and Rescue System in June 2001.

2 : Training

Training is the alpha and omega of search and rescue. Responders begin training years before they have an opportunity to make their first rescue, and they continue to train after every deployment. They train the entire length of their careers, learning everything from how to properly wield a concrete-cutting chainsaw while hanging eighty feet in the air to the best ways to jump into subzero waters during a midnight flood. Once they have mastered these techniques and everything else they can possibly learn about the search and rescue trade, they train some more.

Training props, such as this derailed passenger train, provide extremely realistic scenarios for responders.

Welcome to Disaster City.

Texas Task Force 1 trains with other teams during annual full-scale exercises on training props designed to replicate specific disaster scenarios, such as the 1985 Mexico City earthquake.

Training 11

The initial training area of Disaster City, the Technical Skills Training Area, serves as the primer section of the search and rescue training regimen. This portion of the complex includes disciplines such as lifting and moving, shoring, and breaching and breaking. The smoke in the background is from the Brayton Fire Training Field, Disaster City's neighbor and the largest live-fire training facility in the world. The fire field is under the direction of the Texas Engineering Extension Service, which also serves as the sponsoring agency of Texas Task Force 1.

Texas Task Force 1 trains at Disaster City. This complex has all the equipment responders love working with, like full-sized derailed passenger trains, pancaked parking garages, and decimated strip malls. Ultimately, realism is the main key in training responders to be completely prepared for the unthinkable.

Members begin their education at Disaster City's Technical Skills Training Area—sort of the primary school before heading out to the university level.

▶ Trench rescue is an important component of search and rescue. In this training example, a pneumatic shore is being lowered to support side walls before the responders enter to assist a trapped victim.

▼ Various means of supporting and shoring structures are used in training at Disaster City. In this training exercise, students are learning how to stabilize concrete bridge beams that have crushed a passenger car and school bus.

This part of the city teaches responders about the basic components of search and rescue before putting all the pieces together to be used in a disaster. Just as in preschool, responders begin with blocks. They learn how to move them effectively and how to stack smaller blocks onto larger blocks. Unlike those in preschool, these blocks weigh in excess of twelve thousand pounds, and a team of six men and women may have the liberty of working only with a few metal pipes, so they have to learn the basics of lifting and moving that pertain to search and rescue.

The basic physics of how to lift and move massive objects that have collapsed is of extreme importance in a disaster environment. Whether a responder is attempting to remove a huge beam from a victim's leg or support the load of the three unstable floors above, an intense education is needed to be able to perform effectively while not always having tools within working distance. During

Using a search camera, a responder trains on a twenty thousand–square-foot concrete and rebar rubble pile at Disaster City.

their lifting and moving training, responders start by learning to use their minds and leverage, whether it be with pipes, ropes, or wood—or a combination of all three. As the responders advance in their education, they are granted more luxuries, such as cranes and cables, which may eventually become available at a disaster scene.

Next in the educational quest to become a member of TX-TF1 is studying the more specific tools of the trade. Rescue tools include equipment such as common nail guns and power saws as well as

Two students wearing hazardous-materials protective suits analyze potentially harmful contaminants in front of Disaster City's strip mall complex. The mall features a series of collapsible ceilings and walls.

◀ As cranes and heavy-lifting equipment are not always readily available in the heart of the disaster theater, complex technical skills must be learned. This group of students is learning how to construct and maneuver a wooden bipod to remove a several thousand–pound slab of concrete.

▼ The breaching and breaking section of Disaster City features a series of tunnels and replaceable wood, metal, and concrete panels. Students use everything from crowbars to diamond-chained chainsaws to strategically cut through, or breach, the panels to gain access to victims that may be inside. A surprising number of technical aspects are involved in breaching concrete panels, such as determining the proper shape and size of breach holes and deciding whether the concrete should be breached from a horizontal or vertical perspective. Because of the large number of concrete structures in America, ranging from large box warehouse stores to strip malls, Texas Task Force 1 spends a substantial amount of time training on the principles of breaching/stabilizing concrete and its support materials.

These are the exterior and interior of a single-family dwelling at the Disaster City training complex. This structure has been created to authentically replicate what one would expect to find in someone's home after a tornado. The buildings at Disaster City are fully customizable with movable roofs, walls, and breach panels for authentic training experiences.

"Shoring" creates a series of supports to assist in stabilizing walls, ceilings, and floors. These students are learning to use wood in strategically constructed patterns to support a multi-ton load. Although metal pneumatic shores are often used in search and rescue, responders are trained comprehensively on the use of lumber, since it is so readily available in the United States.

complex listening devices and search cameras. There is a very precise formula for cutting into a concrete wall, and that formula is extremely difficult to follow without knowing how to use the proper gear. When dealing with structural collapses, Texas Task Force 1 uses its tool training primarily for structural shoring and breaking through walls, rooftops, and floors.

Basically, shoring is the supporting of collapsed or unsound structures so that responders can work around and within them while keeping themselves and the victims as safe as possible. Think of shoring as the wooden braces used to support mine shafts. Team members use shores by strategically building a series of supports whose shape and size have been proven to assist with certain loads of weight pressure. The shores vary in their complexity and size depending upon the amount of resistance pressure, and they are built with either pipes or general lumber you can find at any lumberyard. Pneumatic shores are prefabricated adjustable pipes that attach at their end points. While these are quick and relatively simple to assemble, there are certain instances where they cannot be used because of the location of the incident. The responding squad may simply not have the resources to transport the shores to the disaster area, or once the shores are built, it may prove impossible to remove them if the structure shifts. For this reason, responders also become masters in using wood to shore. Lumber is an excellent tool for search and rescue teams, as it is readily available nearly everywhere in America, easily transportable to the site, and expendable if unable to be removed. When foreign responders train, however, they need exposure to both types of shoring because wood is often more difficult to attain in other parts of the world, thus making pneumatic shores a preferred choice in many countries.

As the responders build upon their skills and gain confidence in the Technical Skills Training Area, they learn the most important tool they possess is their ability to bring creative solutions to a disaster environment. Search and rescue responders worldwide have gained the reputation of being "jacks of all trades" and possess the ability to put order to chaos.

After learning basic lifting and moving, tool education, shoring, and breaching and breaking, responders then move to the main training grounds of Disaster City and the real challenges begin. Resembling the set of a motion picture, the landscape is the scene of a major catastrophe—adorned

The light of two headlamps allows responders to secure a victim for transport out of the basement of a Disaster City home.

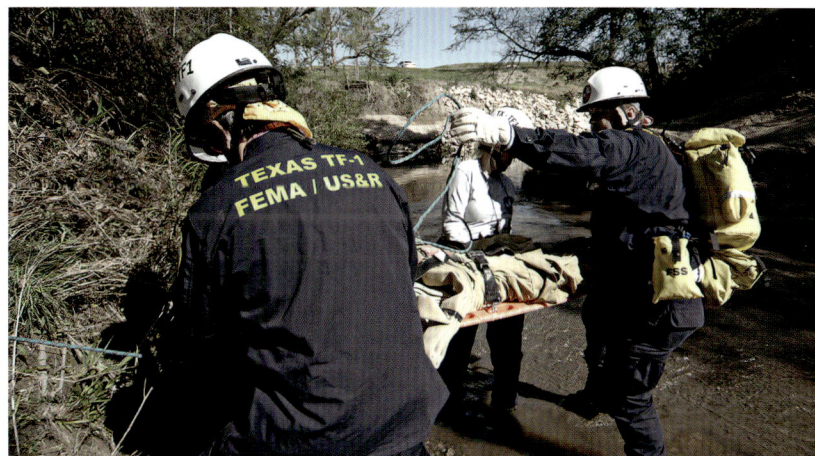

Texas Task Force 1 members remove a victim from the banks of a stream during a Disaster City training exercise. Volunteer victims provide assistance during task force operational exercises and are outfitted with injury makeup and information cards giving details about the injuries and symptoms they are experiencing, providing an authentic representation of a disaster response.

◀ A responder performs high-angle rescue work on a hanging slab of concrete at a Disaster City structure that replicates the aftermath of the Oklahoma City terrorist attack on the Alfred P. Murrah Federal Building. High-angle rescue is an extremely complex, technical form of rescue. In this instance, the responder is prepping the site for a breach through the concrete and rebar panel, allowing access to the other side. Breaches such as this are difficult because the loose concrete must be secured so it cannot fall and injure potential victims below.

with smashed cars, exploded buildings, strewn concrete and wood, collapsed strip malls, and on and on. This miniature training city has most of the infrastructure one would expect to find in Anytown, U.S.A.—the only difference is that it is all wrecked.

Nearly any type of catastrophe replication can be created within the complex. A lake on one side of the city allows for water-rescue training and the staging of victim evacuations. On-site staff prepare the complex before training exercises to create the type of environment needed for a specific deployment's mission.

Even though training is a continual process for the task force, each year the entire team responds to a full-scale exercise modeled as an actual deployment. Once the members have been activated, they have a four-hour window to grab their gear and head to the team's headquarters and staging area, located north of Houston. Once on location and processed, they will travel to Disaster City, set up the base of operations, ready logistics, and begin their search and rescue missions. The members then eat, sleep, and work on-site for three days, exactly as they would during an actual natural disaster or terrorist attack.

▶ Accessing victims can often be a precarious task.

A rescue squad member gives directions about where additional equipment will be needed.

An army of volunteers assists with the full-scale exercises, serving as "victims" during the training exercises. Complete with injury makeup (a combination of a corn syrup and red food coloring mix along with small props), ranging from scratches to protruding bones, the volunteers are placed throughout the complex and given detailed instructions about what injuries they have so they can transfer that information to the responders. The victims play an integral role in bringing a sense of reality to the exercise and challenge the responders to work with real human beings. Interaction with the victims during the high-stress environment of a mock disaster helps prepare the responders to be able to save lives in a real event.

▶ Volunteers play a key role in providing realism to training exercises in Disaster City.

Managers and observer controllers at work during a full-scale training exercise.

Texas Task Force 1 and a team from the United Kingdom work to extricate a victim from the rubble.

Not all responders are human. Here, a search robot maneuvers through a series of obstacles during a training evaluation exercise designed to test various robotic capabilities and how they pertain to search and rescue. The evaluation is organized through a partnership between the Department of Homeland Security, the National Institute of Standards and Technology, and the Texas Engineering Extension Service.

Training 23

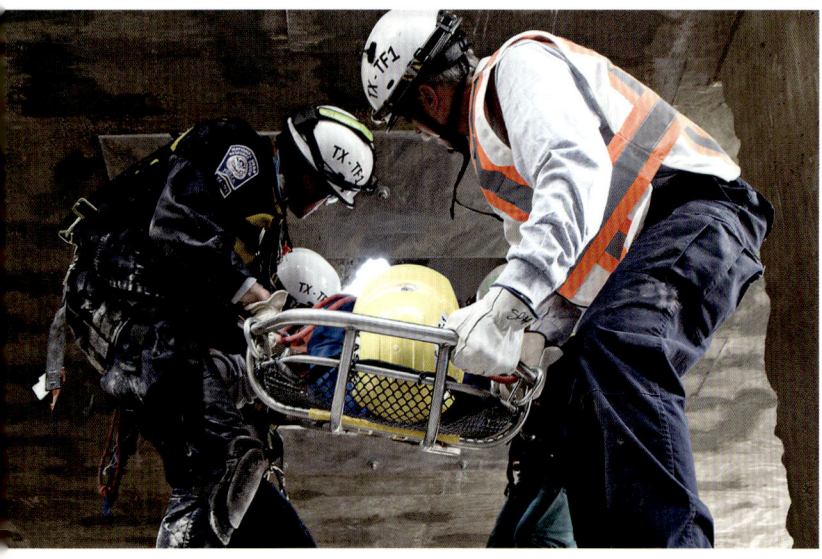

Medics secure a victim into a basket for removal through a hole they created in a concrete wall.

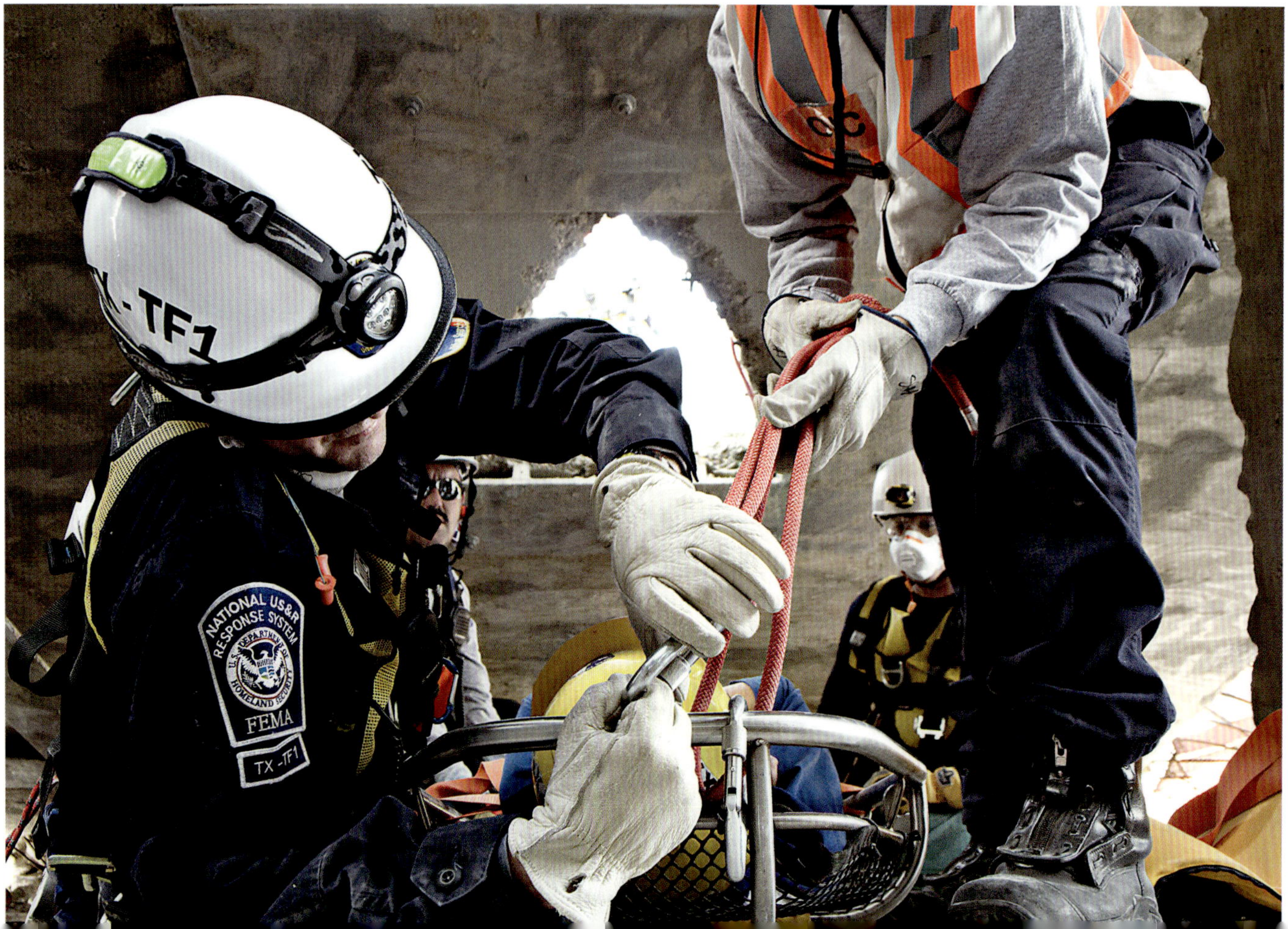

Responders search for victims on a wood rubble pile. This is similar to the type of environment faced after a hurricane or tsunami make landfall.

Anchored support is of crucial importance when performing high-angle rescues.

The full-scale exercises, along with TX-TF1's regular training throughout the year, allow the team to be one of the best-prepared, most efficient urban search and rescue teams and water-strike forces in the United States.

Water-Rescue Training

Texas Task Force 1's water-rescue training is held at areas replicating actual disaster flood events, such as river rapids. Team members learn specialized techniques, from removing victims safely from shorelines to the more advanced skills of maneuvering rapids and avoiding "stringers," or branches, fences, and other debris that may be submersed in the water.

A responder breaks through a concrete floor to gain access to a victim below.

"Evidence" used during a training exercise for a response to an attack by weapons of mass destruction.

American and UK responders work together to stabilize a vehicle and safely remove a victim.

Responders move rubble by hand to access a small break in the concrete foundation.

Responders transport a Disaster City volunteer victim from the scene.

Observer controllers wear vests to make themselves "invisible" during training exercises at Disaster City, allowing them to make recommendations to the team during post-exercise briefings.

A Canadian responder watches as a Texas Task Force 1 member below secures a victim in the rubble.

After breaking through the concrete and rebar during a training exercise, responders save a victim from the collapse below.

3 : Equipment

Search and rescue gear is as diverse as the craft itself. Equipment needs of a swift-water rescue technician vary greatly from those of a frontline medic, and the levels of equipment complexity vary widely. A responder may need to use a twenty-dollar crowbar to remove an impediment so that a quarter-million-dollar search device can squeeze through a crack and get to a child. The equipment used in any given circumstance is contingent on who (or what) is using the gear and what result is needed.

In addition to actual rescue equipment, a large amount of gear is needed for organization and transportation—both in and out of the field. Front-line squad responders may have a bag full of ropes, carabiners, and other high-angle rescue gear, but that gear has to be coordinated via ground, air, and/or water to be systematically available in the field. In addition, a lot of communication and coordination equipment is needed at the incident command level to choreograph an entire search and rescue system to run as a well-oiled machine.

In search and rescue, as in most situations in life, the system structure is only as good as its weakest link. Each responder must be at the top of his or her game in order for the entire system to perform effectively. Every component of equipment must be fully functioning and flawless. An incident responder in the midst of a structural collapse reconnaissance can never be left to wonder if her search camera is going to perform effectively. A swift-water rescue technician cannot afford the few bursts of brain energy to concern himself whether the buckles on his life vest are going to hold. Everyone on the Texas Task Force 1 team has very specific responsibilities, and all attention is expected to be laser-focused on each of those responsibilities. Any concern about

Texas Task Force 1 maintains a $8 million state-of-the-art equipment cache of more than seventy thousand items weighing nearly one-thousand pounds.

Cots line the floor of the Reliant Center in Houston as Texas Task Force 1 prepares to bed down for the night while awaiting Hurricane Gustav's landfall in 2008.

The trucks are loaded before a disaster.

Bags are packed, and team members are ready for deployment.

Logistics provides a pre-daybreak breakfast before heading out into the disaster.

During state deployments, coordination between the task force squads, the local jurisdictions, and the state operations center is always of paramount importance.

◀ A reconnaissance team navigates a coastal neighborhood before a hurricane's landfall to determine low-lying areas and potential areas of evacuation.

Equipment 33

A forward team unloads supplies to ride out Hurricane Alex along the Gulf Coast.

All task force members are trained in land navigation with global positioning system (GPS) units.

A team safely transports a communications satellite.

◀ A forward team updates the base of operations.

▶ Various victim stabilization devices are used in the field.

A responder uses a search camera to see within the void left by this concrete foundation.

Search and rescue gear ranges from the very complex to simple, such as this common reciprocating saw.

◀ Every member of Texas Task Force 1 carries breathing apparatus in his or her gear.

basic necessities, such as where he or she is going to eat or sleep, can take valuable time and energy from a responder whose sole mission is the wide-area search of a sunken city, such as New Orleans after Hurricane Katrina. Therefore, the responsibility for these basic necessities lies with TX-TF1's logistics personnel. If any situation demands gear, shelter, bedding, or food, these men and women provide it, maintain it, fix it, or get it in the field. Logistics is the supremely important foundation of search and rescue—without it, the entire system would not just fail; it would never be able to perform in the first place.

Various pieces of equipment are used to prepare the victim for extrication during a training event.

▶ LED lights on the rod of a search camera allow these training responders to see within a pitch-black void.

Equipment 37

◀ When available, heavy machinery is often the equipment of choice for moving large loads of rubble and shoring materials.

▶ While Texas Task Force 1 was on response near Galveston Island's Crystal Beach, communication to the base of operations was often limited during the first few hours after Hurricane Ike's landfall.

▶▼ Miscellaneous diesel equipment was used on Bolivar Peninsula to light up a squad's overnight camp. The squad had traveled via Blackhawk helicopter to the peninsula the day after Hurricane Ike's landfall and did not yet have access to generators for powering floodlights.

▼ Texas Military Forces work in tandem with Texas Task Force 1 to provide air transport and rescue support.

Squad managers analyze a map of New Orleans while performing reconnaissance in a northeastern sector to search for survivors of Hurricane Katrina.

Dry suits are personal protective full-body suits that are completely waterproof and impermeable to most contaminants. All responding members used these suits when wading through the flooded streets and structures of New Orleans after levees broke around the city in 2005. Although the suits do not let moisture in, they also do not let it out, creating pools of perspiration inside. Each returning squad must thoroughly wash the suits to keep them in prime condition.

Satellite communication systems are present at the base of operations at many disasters.

◀ While in the field, Texas Task Force 1 usually sleeps in tents such as this one being constructed. Customizable, the tents can be converted to be winter or summer specific, outfitted with heaters, insulated nylon walls, and even fans and air-conditioning units.

▶ Contamination is generally a major concern during flooding events. Therefore, hazardous-materials decontamination stations are created for responders returning from the field. The decontamination stations can range from high-pressure washing areas, like a car wash, to simple bucket scrub-down areas, such as this one. Responders file through a station line where they receive a different level of decontamination at each station. Once complete, they can then return to the general population of their squad and the base of operations.

During disaster response events, Texas Task Force 1 is often staged with other federal task forces, assistance personnel, and military. The agencies and organizations eat, sleep, and work together during deployments.

An equipment cache expansive enough to supply Texas Task Force 1 for up to ten days is shipped by either a fleet of trucks or cargo military aircraft, depending on the type of deployment and location.

The Tools of the Trade

Hydraulic concrete chain saw with diamond chain

Search camera with telescoping search and rescue camera system

Exothermic torch

Air-lifting bag system, capable of inflating and lifting several tons

Hydraulic fourteen-inch concrete/steel cutoff saw

Electric and cordless wood-cutting circular saws

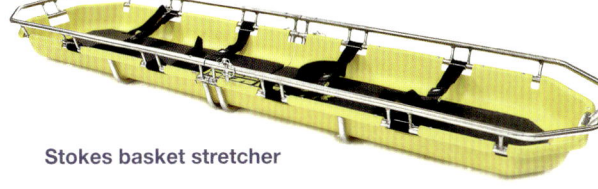

Stokes basket stretcher

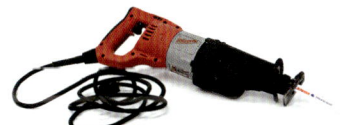

Electric reciprocating saw

Hydraulic concrete hammer drill with bit

Electric rotary hammer with bit

Concrete coring drill with diamond bit

Cordless steel rebar cutter

Hydraulic concrete breaker

Equipment 43

Gasoline-powered concrete/
steel cutoff saw

Delsar Life Detector seismic/
acoustic electronic listening
device

Thermal-imaging camera

Chemical warfare–
agent detector

Sixty-inch pinch/
pry bar

Air sampler

MRE (Meal, Ready-to-Eat)

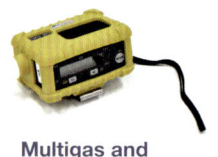

Multigas and
chemical detector

Defibrillator

Flotation device

Gamma multigas and
radiation detector

Fire axe

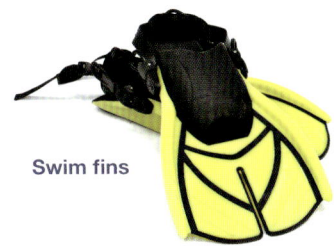
Swim fins

One thousand–watt suitcase generator

Search and rescue helmet

River board

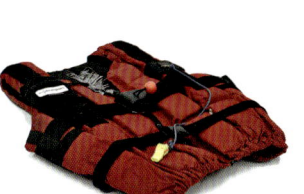
PFD (personal flotation device)

Rescue Rocket line-throwing devic

Water-rescue throw bag with seventy-five-foot rope

Water-rescue helmet

Smoke marker, sea dye marker, handheld flare

Quick-hoist collar

Equipment 45

Equipment 47

Responder Interview

Responder: Chuck Jones
Position: Task force leader—white team; former logistics manager
Search and rescue career: 35 years

Q: What is your history with Texas Task Force 1?
A: I applied and was accepted as a logistics manager on the task force with the first wave of people in 1997. I remained in that position for twelve years, and I have been on every major deployment the task force has responded to since. Within the last few years, I was promoted to task force leader of the white team. I have thirty-five total years of duty with the fire and rescue service in general, beginning as a teenager with a volunteer department. I hired on as a career firefighter in 1979 and have been with the same fire department since.

Q: What exactly is logistics as it relates to search and rescue?
A: Search and rescue logistics is responsible for the movement, housing, and support of the team—the feeding and the well-being of all the responders in the field. From the standpoint of "what does logistics do?" it's the backbone of the task force. It's getting items as simple as lip balm sometimes. Where do you get lip balm when there's no grocery store? The logistics folks know how to do that; they know the little techniques to get stuff like that done quickly. Some of that may not seem overly important until you're the one who needs it. Making sure someone has those basic things is fundamental for the survival of the rescuer—if the rescuers aren't healthy, then they cannot do their job.

Q: Why did you begin in logistics with the task force?
A: I felt like logistics was going to be a good fit with me because I enjoy working with my hands, but I also enjoy working technically. I like trying to understand people explaining a situation and then having to devise a solution or a tool to help them achieve what they're trying to do.

Q: How did you, serving as the task force's logistical manager in the beginning, create the initial operational program?
A: We came out of the gate training and getting a list of initial equipment together based on what the current federal teams were using at the time. We looked at what they had, and we purchased all the same gear. When I came on, one of my first jobs was to find other logistics specialists. It was more about trying to find people who could work with their hands and think on their feet than it was people trained in logistics. At that time, we were solely relying on fire department and public safety personnel because those types of people had an understanding about what could be expected during a disaster situation and would not shy away.

We spent the first couple years training and getting our feet wet. We initially learned a lot from the forest service by taking the same courses that the wildland firefighters were taking. Those firefighters are great at supporting four thousand people at a base camp and trying to feed them, clean them, and supply their equipment. That was where we really learned how to support and move the task force.

But we also reached out to the military to assist us because we assumed we would always be moving fast, and moving fast must be done via aircraft. They came down and helped us format our cache of equipment onto military-styled pallets that could be moved by military aircraft like C-130s. That was the first packaging system we used, and we still use it today.

When we first started, we didn't have the amount of funding that some of the federal task forces had, so we used our ingenuity for some of the things we did. For instance, we couldn't buy actual field sinks, so we bought some household stainless-steel kitchen sinks and mounted them inside plastic trunks for both transport and use. There were a lot of those types of things back then. We slept in the single fire station bay where we were based to try to get all this stuff ready to go.

The day we finally received the approval to get inspected by the federal system with the possibility of getting into the system, we were all extremely excited and nervous. We took all the equipment and laid it out throughout the entire station. We had heard another team had the quickest inspection record set at a day and a half, and we wanted to beat that and really impress everyone. We were determined to beat it and to pass quicker than anyone ever had in the system—even though we weren't actually even in the system yet. We laid each piece out on the floor in its respective order on the inspection sheet. That way, the inspectors could start at the top and inspect the "A" piece of equipment and so on all the way through. All the equipment was laid out row after row after row. Everything had to be inspected regarding when it was purchased to where it was purchased to how it was purchased. The government had to know where all the funding came from. It took them two and a half hours to inspect us, and we were done and we passed. We had everything we needed—not to mention we had beaten the quickest federal team at the time by more than thirty hours of inspection time.

Back in those days, we personally inspected and moved the equipment so much we all knew where each piece individually went in each case and on what shelf. We would learn what equipment we liked in which box because it would be better suited. There eventually came a time when all of us were so familiar with all the equipment that if you asked where something was, no one had to go to a database or look at a piece of paper but could say, "It's on pallet four, on the third row, in box six." Everyone was truly that familiar with it.

We felt really good about that. The more we know about our equipment and exactly how the responders will use it out in the field, the easier it is to support them. It's unbelievable to look back at it now. Back then, there were just a few of us and that was it—it completely blows your mind to see it today. The facilities are so phenomenal. All of our equipment is organized in an electronic, searchable database that has items, keywords, inventory, and descriptions. Today, we have three times the equipment we did when we started. It's more than we ever would have expected.

Q: What is an average day on deployment for a logistics specialist?
A: It's a continuation of making sure the people who are going to go out and do the work and make the rescues have all the necessary equipment they need to do their job during that operational cycle. Normally, when a cycle is about to begin, logistics is approximately two hours ahead of everyone else so that we're prepared in the beginning when everyone else is ready to go. We send logistics specialists directly into the field with resupply equipment to support the group that is going out the door. While

they're out there, we'll have another logistics team at the base of operations who are replenishing the supplies that were just sent out on the previous operation—making sure the teams have enough fuel and that the repairs are being made. If we cannot make the repairs ourselves, then it's logistics' responsibility to find a place where it can get done.

Q: So experience certainly plays a primary role?
A: Experience is important, but it's a tight combination of experience and training. Experience plays a key role in knowing all the details. For instance, you learn that if you have a cooler with ice water in it and that all the people who come by are going to stick their dirty hands in it to get some bottled water, then you're going to need bleach or some type of antibacterial protection in that ice. You learn that if you miss simple things like that, you can put an entire group at risk and out of commission. It's the lessons learned and the sharing of those lessons—the training. Everyone shares information with everyone. People get that information through both the training and experience.

Q: What are some of your most memorable logistical field operations?
A: One of my vivid personal memories was after Hurricane Katrina in New Orleans. It was sweltering hot, and mosquitoes were everywhere. Everyone was sweating, and there was no running water. We were three or four days into the deployment, and we could see from the logistics side that the responders needed something special—something that would lift their spirits. As meaningless as it may sound, we gave them each two 16-ounce bottles of water for a "shower." We set up tarps and tents and all the stuff they'd need to go in and take a shower with their two bottles of water. That way they could get off some of the grunge and the smell and the nastiness of the disaster. I think if you ask people who were on that deployment at that time what they remember getting when they got back to that camp, it would probably be those water-bottle showers. That really lifted their spirits, and that's what you have to do. You have to make sure the personnel are taken care of, because if they're not, then they cannot do their job. The little extras like water-bottle showers always seem to help a lot.

Another time was at Ground Zero in New York City. We had been there for several days and were hauling debris day after day and hour after hour but not finding anyone. Needless to say, depression began setting in. So one of the members got in touch with a contact at Blue Bell ice cream in Texas. Blue Bell then shipped two huge boxes of ice cream on dry ice so we could give those to the responders when they came in off their shift. Logistical operations can be so extremely complex and we can manage such large-scale operations, but the little things like water-bottle showers and small cups of ice cream stick out in my mind from a logistical operations standpoint.

Q: Has there ever been a time when the logistics staff has not been able to supply what the task force has needed?
A: No, never. You must be resourceful. There has never been a time when we haven't been able to supply the team with what they needed. We might not have been able to supply them with what they wanted, but we have always supplied whatever they needed. At the end of the day, logistics makes sure our responders have their necessities.

4 : Responders

Getting the Call

While in the midst of any usual activity on any given day, a responder receives a phone call from the automated response program that signals each member of Texas Task Force 1 that he or she is being called to duty. He types in his specific response code to let the system know that he is available and ready, and then he leaves everything he is doing and immediately heads to the team's staging area north of Houston.

Once on location, he and the other task force members begin response processing, which includes gear checks, physicals, organizational paperwork, and initial disaster event briefings. After processing is complete, he leaves with his team to the disaster venue (in this case, say, a major earthquake) either by task force vehicle convoy or military aircraft. As he arrives, he is briefed on the situation at the field base of operations and then assigned to his smaller squad, depending on his task force position.

This responder, for the purpose of example, is a structural collapse specialist. He leaves with his squad to cover a predetermined section of the disaster—the northeast side of a collapsed, or "pancaked," multistory parking garage. Intelligence reports show there may be a live victim in the area and the structure is extremely unstable. Upon arriving on scene, the responder and a teammate pull out a series of cables and small metal cylinders. Attaching a cable spiderweb throughout each of the various cylinders, he strategically lays the contraption around the perimeter of the garage. The responder has just created a hypersensitive listening device capable of allowing him to hear the scrape of a fingernail on concrete from more than a quarter mile away.

After the technical setup is complete, his teammate hears the faint whisper of someone saying "please." Although he has a general idea of the victim's location from the device, his squad's canine

Anthony Tortorice

Breaching and breaking.

handler and her specially trained search dog will pinpoint the location. The handler instructs the dog to use its nose and ears to search the predetermined area, locate the victim, and then bark for at least thirty seconds to signal the person's location. Within minutes, she discovers a young woman wedged at least ten feet down a tight cavern of collapsed garage columns. The responder pulls out a specialty search camera to judge the amount of instability in the cavern. Gingerly maneuvering the camera's shaft throughout the small concrete opening, he is able to tell that he will have to shore, or add support, to the cavern once he has breached a hole that is wide enough to gain access through the front.

"I need a concrete saw," he says to his teammate as he begins preparing to cut a hole through the slab of concrete foundation that blocks the way.

But, because he is above the victim, he cannot just cut through the concrete; it would fall on her as soon as it was free from the main foundation. Instead, he drills two holes through the slab and loops a piece of nylon rope inside one hole and out the other. This way, his teammate can hold the concrete block with the looped rope once it has been freed, then lift it out of the way without harming the woman below. With his saw, a diamond-bladed circular saw on steroids, he cuts a large triangle through the concrete slab and metal rebar that is large enough for him to maneuver his body through.

After successfully removing the block, he shines his headlamp on the woman's face and asks her to remain calm. "I'll have you out quickly," he says. The responder's teammate then hands him a series of metal pipes that he begins inserting throughout the cavern as he makes his way down. These pipes are actually pneumatic shores that support massive loads without collapsing. Similar to the wooden frames built in mining tunnels, these shores will keep both the responder and the woman from being crushed in the rubble in case of any shifting in the massive load above. Within minutes, he is more than ten feet under the earth. He gives the woman a quick injury analysis and secures her to a stretcher, which is hoisted out of the cavern by his teammates. Finally, the woman is free from the rubble and being treated by the medics. The responder radios the squad's progress back to the base of operations, and they then instruct him of his next course of action to continue saving victims' lives.

Hours later, when the responder's shift rotation is complete, a fresh squad relieves his duty and he returns to the camp that the logistics staff has built, eats an MRE, or "Meal, Ready-to-Eat," and then gets some quick rest before his next shift begins. With complete focus, he continues this type of routine for the next ten days until the task force is demobilized. Once the demobilization orders have been passed, the responder and the rest of the team are transported back to the staging location, where they are debriefed and receive additional physicals. From there, he will either return home or be issued at least a full day's rest before returning to deployment. Once he finally does get back home, he eagerly awaits the next time he is signaled for duty.

This is an authentic representation of what it is like to work in search and rescue. It is not human instinct to go into an unstable rubble pile that was once a parking garage; instinct is to run from it. What is it that makes some people leave a comfortable life and family to go crawl into a concrete kaleidoscope with the sole focus of saving someone else's life while risking their own?

Some of us may have wanted to grow up to be firefighters, police officers, soldiers, or explorers; as we mature and become older, most of us begin to limit our physical risks on a day-to-day level. Any thoughts of being a physical "hero" subside as we learn more about life and all it has to offer. We move away from our heroic childhood role models, and we learn that one does not have to be a comic book superhero to help others. Our responsibilities to ourselves and our families increase, and the natural thought process of minimizing risk for ourselves and our loved ones takes hold somewhere along the way. Most people continue to want to help their fellow human beings—and do—in a variety of ways, whether giving to charity, volunteering, teaching, or carrying out some other community service.

Dr. Alexandre Migala reviews medical needs of medical special needs patients before evacuating the patients via military aircraft in preparation for the landfall of Hurricane Gustav.

Responders often face extreme temperatures and harsh conditions on deployment.

Working in tight spaces is usually part of the job.

Respiratory protection is required when working in contaminated environments.

Responders train in the mock-up of a home's basement.

Texas Task Force 1 travels via bus, convoy, or military aircraft (as seen here) depending on the scope of the disaster and its location.

Collection of gear and medical processing take place before deployment.

Responder Michael Cockrell heads back to the team after discussing air operations with Texas Military Forces.

Responders 59

Although the task force's primary mission is the search and rescue of human victims, responders do what they can to assist victims' four-legged companions.

Some people, however, never lose a deeply rooted desire to physically help others in danger while simultaneously receiving the adrenaline rush of doing so. This type of spirit defines the men and women who work in search and rescue. In choosing the career of emergency response, whether in medical assistance, firefighting, or search and rescue, one automatically adds a number of mental and physical extremes into one's life. Once someone makes the career choice, that individual has officially chosen to allow both mentally and physically harmful situations—car wrecks, hurricanes, terrorist attacks—into his or her life on a regular basis and has *volunteered* for a front-row seat.

So, why would someone scale scores of flights of stairs in a building that he knows is about to collapse

Ken Larsen assists evacuees onto a Blackhawk helicopter for transport to safety.

Task force members return to Galveston Island after forty-eight hours of rescues on nearby Crystal Beach.

◄ Two evacuees are transported to a safety zone after being rescued from their bed-and-breakfast inn on Crystal Beach. Found two days after Hurricane Ike made landfall, the women had scraped "Please Help Us: Supplies Needed" on the beach. Their inn was one of the few structures on Crystal Beach that survived.

Two responders briefly rest while awaiting air transport to Crystal Beach during the Hurricane Ike response in 2008.

Outfitted with completely waterproof and contaminant-resistant dry suits, water-strike team personnel wade through the flooded streets of Galveston Island only hours after Hurricane Ike's landfall.

A responder searches a structure after suspecting there is someone inside.

64 CHAPTER 4

in the slim hope of saving one other person's life, a person he does not know, even when he realizes that he, himself, might die? Since he does not know the victim, he also does not know the person's personality in daily life—whether the person is a great parent or a criminal—he knows nothing about the victim. But he is still willing to throw his life on the line with complete selflessness to save that other human being. Why? This may be the question most asked by news agencies and reporters who have been on the scene with Texas Task Force 1.

When the task force was responding to the aftermath of Hurricane Katrina's wrath in New Orleans, a responder told me that he always pictured his mother as being the victim in need. Whether it was man or woman, young or old, he always related the victim's needs to that of his mom as if she were in the same situation. But whatever his personal reasons, he felt it was his duty to help others in situations where they could not help themselves.

The scenarios in which one may physically need to help someone else in danger lead to intense high-stress situations—the types of situations that pump adrenaline. No search and rescue technician wants a hurricane to come ashore and cause millions of dollars in damage and crush the lives of so many. Nobody wants to witness the despair of a major flood, earthquake, or terrorist attack. But if it happens—and it does and will—responders want to be the first people in there. Most of them have learned to use the rush of adrenaline to their benefit and have learned to live on it. In fact, they embrace it.

Responders channel the emotions and adrenaline that come with their jobs into a single focus on saving others, even while putting themselves at risk—especially while putting themselves at risk.

Responder Mike Cockrell receives information from a resident about other survivors that may be present on Bolivar Peninsula after Hurricane Ike's wrath.

Texas Task Force 1 member Ken Larsen awaits the arrival of an aircraft transport to Bolivar Peninsula.

Hot temperatures persisted in downtown New Orleans after Hurricane Katrina. Old banner advertisements were strung as tarps to provide shade to evacuees.

Canine handler Bob Deeds travels in the back of military transport to the pine forests of East Texas in search of trapped victims.

A responder gets a big thank-you from a flooding victim.

Jim Yeager, Texas Task Force 1 search manager, speaks to the field base of operations via satellite phone while traveling into downtown New Orleans for search operations after Hurricane Katrina.

Why They Do It

RESPONDER: SUSANN BROWN

"I recognize that any of the events I have been to could have happened to me or someone I care deeply about—especially with the reality of hurricane response and living in Texas. I realize that any of these events could potentially affect me on a very personal level. It's a service; I want to give back; I've had a good life, and I feel like I owe for that good life. This is one of the ways I pay that back by trying to make it a little bit better in a situation that is really awful. Sometimes I may have a lot of impact and sometimes I may have little impact, but if I can make it just a little bit better, then I feel like I have paid something back for the life's blessings I've had. But we are all different in our motivation. Some people are out there working for completely different reasons, but they're still out there working as hard as they can and that's what matters. When it comes down to actual motivation, it doesn't really matter—as long as you are in fact motivated and making a positive difference. It's such an honor to work with people who have this mind-set and to be able to call them my friends. There's a feeling in the room when I walk in and I see the faces of the other responders I work with. My stress level drops because I know that whatever happens, we'll figure it out and do what we need to do to get the job done. I know that because I know the people in that room can do it."

RESPONDER: MATTHEW MINSON

"To put it on the most basic level, I believe the common reason is heroism. But I think there's more to it than that. There's a common quality in character that is a little more profound than such a simple term. I think if we were to break it down completely, the thought process of responders willing to give their lives is to actually preserve the greater good of humanity at the cost of one's own life. This is as inherent as their genetic code.

The principle of Texas Task Force 1 and all the other task forces out there is the same principle that allows people to run into a collapsing building to save others at their own expense. The principle of helping others is as fundamental to the search and rescue members I know as is breathing."

RESPONDER: CHUCK JONES

"We all do it—everyone in this—do it because we want to help others, as cliché as that sounds. People generally dial 911 when they run out of ideas. For whatever the reason, they can no longer handle the situation. At that point, they need someone to come in and solve that specific problem—to get them out of that problem. Part of being a responder is solving those problems. It's a horrible thing to see someone's possessions destroyed in a storm and to see people hurting—it's terrible. But, it's an adrenaline rush to go help them, and responders are adrenaline junkies. It's what they live on. The ability to help others while increasing that adrenaline. All of those responders who have responded time and time again, day after day, look at the inconveniences and say it's not a big deal, and then they go out there and do it again. If it's going to happen and I can't change it and I can't stop it from being devastating and entrapping people, then I can sure as hell be the first one there to help. And that's what every member on this team and all the teams across America feel about it. Period."

RESPONDER: BILLY PARKER

"I think it begins with my faith. I think it's important for everyone to have a mission in life. For me, I feel like that mission is search and rescue. I have thought the same thing since my high school days. Whether serving as a Big Brother or Big Sister or simply helping your fellow man, there's a connection in the emergency response field. There's a distinct camaraderie there that links these types of feelings. Emergency responders lay their lives on the line, sometimes on a daily basis. This isn't for glory per se, but for the sole reason of sincerely helping that

Task force leader Billy Parker speaks with a Hurricane Katrina victim who was rescued after being stranded in his home for several days.

other person and getting the reward of positive feelings that come from doing so. Emergency response and search and rescue is a distinct path where you are able to do just that."

Responder Interview

Responder: Matthew Minson
Position: Team doctor
Search and rescue career: 11 years

Q: When did you decide you wanted to get into medicine and help others?
A: I got into medicine because I saw a movie when I was about five years old, and there's this scene where these lepers are ridiculed and just treated terribly by people. This really touched me, and I started crying about it while I was watching. My mom came out and asked what was bothering me. I told her I was upset about the lepers, and she told me I should try to do something about it—maybe I should become a doctor.

To a five-year-old, that seemed like a very logical answer. So, I always kind of had this interest in medicine, mostly just so I could help people. I was really poor, but I had the good fortune to get a job mopping the floor in a hospital when I was young, and I got a chance to see excellent doctors who took care of their patients. They sort of became my role models. One thing led to another and progressed through a bunch of jobs in the hospital, including cafeteria dishwasher, surgical technician, and paramedic. I did a lot of stuff that gave me a lot of contacts well before I went into medical school. When I got out,

I just felt so grateful. I realized that nowhere else outside America could someone be born into the socioeconomic class I was and be provided with the opportunities that I received and the career I have.

Q: How did you initially get involved with Texas Task Force 1?
A: It stems back to my need to give back for what I've been blessed with in life. When I was formerly an emergency services medical director, one of my medics was a member of the task force, and he explained to me what they did and how they responded to disasters. I thought it sounded like a perfect idea, and it met that desire I had to provide some level of service back. I explored it further, found out they needed doctors on the team, and applied. My first deployment was Tropical Storm Allison.

Prior to Allison, I had been going to all this bizarre training where they were putting me down tiny little holes and popping me into boats and hanging me off buildings, and I thought, "What in the world have I gotten into?" But when everything came together under disaster conditions, it became incredibly impressive how logical the training was.

I joined Texas Task Force 1 because I honestly wanted to do something to contribute back to the country. I would have never had the opportunity to do what I do if I had been born anyplace else. In this country, if you're willing to work really hard, you have the opportunity to do it. And so, I feel people are morally obligated to give back. The task force is a unique opportunity to do that.

Q: You mention a variety of training. What does your cross training include?

A: Structural-collapse rescue, high-angle rope rescue, confined-space rescue, swift-water rescue, and I'm a hazardous-materials technician. Professionally, I'm part of the tactical medical community and have extensive training in public health.

Q: What has been your most challenging deployment?
A: There hasn't been a "most challenging deployment." But there have been three main deployments that stick out in my mind because of the ways in which they were challenging.

PROFESSIONAL CHALLENGES
One of my first major deployments was to Ground Zero after the World Trade Center attacks in New York City. That event was defining in a lot of ways. It was defining for the country, defining for search and rescue, and defining on a personal level.

Ground Zero was so isolating and difficult on a professional level in that there wasn't a lot of information on the current environmental risks. Inherently, it seemed to me there were some severe risks, and, because I was still a bit of a neophyte in the disaster medicine arena, I had some doubts about my current ability. But I used my logic and the fact that I knew my job was to take care of our team as the foundation for my work. My mission was very poignant in the case that I needed to make sure our folks were safe. So there was a huge preventive element with medicine that came into play while at that event.

Some people didn't want to wear respiratory protective equipment because the breathing environment appeared to them to be normal at times. But I was fairly resolute about it and dug in my heels— even when there was an absence of information sometimes—to keep people disciplined in regard to that application. As I look back now, I think of it as one of my finer hours on Texas Task Force 1 because it was a somewhat unpopular position I had taken.

It would have been much easier if I had been a "good guy" and let all my team do what they pleased. Because of the way that search and rescue responders feel about their missions—the distinguishably honorable way they want to go and go and go to accomplish the mission at hand—I and the other team doctors sometimes have to say, "Hang on. Let's make sure you're protecting yourself medically to the best of your ability before going in and trying to save everybody."

Once the Ground Zero response was finished and the information came out that there were specific environmental hazards from the destruction's airborne pathogens, it verified we had done the right thing.

PHYSICAL CHALLENGES

Later on, in 2005, we were deployed to New Orleans for Hurricane Katrina. Things were more dangerous for us on a more tangible and personal level. You could see it firsthand, and you knew what you were getting into. There was everything from environmental concerns to water to open gas lines to downed electrical lines to extreme social disorder. I think that despite the challenges, the task force shined during this event. I saw my teammates saving people and doing things hourly where they put themselves at substantial risk for their fellow man. And I think that's the zenith of what we're after and the philosophy of the team.

During Katrina, we had been in position roughly forty hours prior to landfall to the storm's northwest, so we could respond very quickly. We were configured for heavy collapse and search, and what we ended up in was a massive water operation.

Of course, there was a tremendous amount of water. But you would come across spots of land that would rise out and make impromptu islands. You had overpasses and car ramps that would emerge from the water onto these islands. People covered the islands in order to get out of the water, and they were in need of rescue.

A large multistory structure that was to the east of a city park was just filled with victims on the second and third stories. There was a mix of people who had exposure issues. A lot of them had horribly exacerbated medical issues, and because of the stress and environment, they were extremely sick. There were issues of folks who were desperate and afraid, and their behavior reflected that. And there were lots of children. We had to triage and consolidate people to determine who we needed to get out immediately for life-saving measures. I and a couple of rescue technicians set up this disaster triage station on a nearby overpass that was jutting out of the water.

Texas Task Force 1 maintains a full cache of medical equipment, but we're configured to be a light and nimble entity, and we're not structured to carry enough medical supplies for mass care. The task force is a search and rescue team and not a mobile hospital in any way. Our supplies are primarily for caring for injuries sustained by the team members themselves out in the field. Basically, our medical supplies allow us to be self-sufficient.

Because of that, the limited medical supplies we had with us had to be prioritized to do the maximum good. Getting people out of there by helicopter was going to be very difficult because there was nowhere for helos to land. If we were going to evacuate people by helicopter, they were going to have to go up on a cable basket or harness. While the Coast Guard was splendid, there's always a potential risk in those types of operations, and that risk has to be built into the equation.

It was very challenging medically as to how the triage system was going to be configured. On that overpass in New Orleans that day, the disaster triage system was quite profound and lasted until everyone had received attention. As time went on, more and more and more people were trying to get to the spot where we were. I estimate the total number of people that day between seven hundred and eight hundred, and all they wanted was to get out. Some could barely get there, but they kept coming. It was an arduous and long process. You think of disaster triage as getting people off the site based on medical needs. In this case though, there were situations where individuals were posing as potential threats. Unlike a regular triage event, disaster triage maintains that you have to remove the threats first, and then the victims based on medical needs. What you're really looking at is a semblance of preventive measures. If there's an altercation or problem, you can expand the number of people who are injured significantly. The purpose of disaster triage is to keep that from occurring. The last thing we want is for something like that to make the situation much worse.

We assisted every victim at that location that day and provided responsible handoffs to safety via boats and helicopters. That event remains one of my most acute memories about Katrina to this day.

A few days later, we received a call about a trapped woman who needed medical assistance. We were already out searching a hospital that was reported to have patients still remaining inside, and I was in attendance in order to have a doctor present for any medical needs that might arise. We searched the hospital, and it turned out that it had indeed been evacuated and there were no victims.

We were on our way back and learned of the lady via radio. She had been trapped in her home for days and was clutching a couch cushion in the water after her husband was killed by the flood. Another task force squad had heard her call out. They shut down their engines and listened to try to pinpoint her location. When they found her, they realized she needed immediate medical attention, and that's when they called us.

The squad already on scene was able to get her out of the house and into a boat. A few minutes later, we met up at this slab of concrete coming up out of the water. By the time I got there, she had lost consciousness and had a very faint pulse. I had to get an IV into her immediately. She was not in good condition; some of her skin had begun to macerate severely, and it made getting the IV line to her very tricky. I was able to hit a line on one of her veins and "commando-taped" the IV to where it would be difficult to remove, since that was basically now her only lifeline.

Even though we'd gotten the fluid in her, I realized she probably was not going to survive a long boat ride to safety, and we had to get a chopper. The rescuers got on the radio, and we had one there immediately, but, as usual, there was nowhere for it to land. A rescue swimmer jumped out, and I turned to him and said that we had to get her out of there ASAP. It would be life threatening to risk her IV coming out and there were no medical personnel on the helicopter, so I was going to have to go with her. I cabled up to the helicopter with the woman, and I had a second liter of fluid in her en route. She came to in the helicopter and ended up surviving and doing just fine. It was such a real gift and so gratifying in the midst of such a horrible situation to be witness to something like that.

That's just one story, but there were countless situations like this that were being performed by not just Texas Task Force 1 but all the responders around the city. You feel really honored to be part of people doing good like that.

EMOTIONAL AND LOGISTICAL CHALLENGES

Every disaster we've been to has been challenging in some fashion, but the Space Shuttle Columbia disaster for me was really difficult, as I knew one of the astronauts because we had been in sports together in my youth. That was a real personal application to the shuttle recovery that made it quite challenging emotionally.

The challenges of the shuttle recovery overall taught us as a nation the logistical needs that must accompany federal responses because of the impact on local infrastructures. After the recovery, you saw the maturing of emergency support functions. This disaster had a profound impact on what those functions could and should and would do in order to help responses. Even though the Columbia disaster was detrimental on so many levels, it served as a great tool that has helped search responses nationwide.

Q: Firefighters on deployment are still working relatively within their usual environment. But doctors on deployment are transferred from working in an extremely sterile hospital with specialized equipment at their disposal to a potentially contaminated environment with no electricity. Speaking from your professional view as a doctor, are you out of your comfort zone when in the field at a disaster?
A: Well, as you said, when working in disaster medicine, which is really what you're practicing in the field, you don't have the luxury or capacity that you are used to having on a normal or day-to-day basis in a hospital. It just doesn't work to re-create the emergency department in the field. It doesn't necessarily take you out of your comfort zone; it just makes you look at the playing field differently. The most attractive and daunting and intimidating thing about the disaster environment is that you have to be functionally and quickly innovative under duress. At times, that's hard to do, but you adapt and overcome.

This makes it difficult on certain levels to speak with peers about specific events because it's such a shift in the paradigm of how you interact with patients, what you do with them, and how quick the experience can be.

Whereas normally you would love to go get a splint you could form to someone's leg, you might actually have to utilize the materials at hand. It's not often that you have to move a large number of patients by boat, yet it happens. It's not often that you have to create a triage station under an overpass—or over an overpass—but sometimes that's what you have to do. There are a thousand examples of that.

Q: What is in the Texas Task Force 1 medical kit?
A: As I stated earlier, we carry enough in our medical cache to basically be self-sustained. What we would never want is to put any type of stress on

the local infrastructure. Although we carry a "light" amount of gear, none of this is a shortcoming. It's that way by design based on what search and rescue is.

Most of our gear remains at the base of operations, which will usually be located on the perimeter of a disaster location. When in the field, we can be strung out for miles, so it can actually get down to what you can carry on your back. Even further, if you have to climb into a hole or go into a void space, you may be forced to take even fewer medical supplies.

So, what that lends itself to is a much better ability of field clinical assessment in evaluating our patients. Even though we have monitors and the gear we need, sometimes the physical limitations or the environment—the remoteness of the situation or the lack of transportation—limits us to carrying a minimal medical pack, which is reduced to some IV support capability, basic monitoring, and the ability to perform foundation management of breathing and circulation.

There are always exceptions, but generally the farther forward you go, the less you are going to have at your disposal.

Q: You have talked a lot about disaster triage. Can you get into some more details about the fundamental components of triage?
A: Triage on its most basic level is sorting patients. It's prioritizing based on need and the potential outcome for a large number of people. It would be great if there were a hard-and-fast rule you followed for all situations. But what you would do for a situation involving terrorism and mass trauma versus a situation in which you're looking at a lot of infectious disease versus a situation in which there's a potentially encumbered environment is all different. These scenarios are going to dictate how you are going to be able to adapt the concepts of triage.

Triage in a "normal" environment would be giving the most dedicated resources to the person who has the worst problem. In a disaster or combat matrix, it's sometimes an inversion of regular triage, where your priority lies with the maximum good for the greatest number of people. That sometimes leads to things that are much different from what we would do on a normal basis. There must be social elements built into the disaster triage formula. That being said, all the elements are still being scientifically and medically made for the greater good of the people.

Q: What are you looking for when analyzing the health of responders in the field?
A: If we could prevent any and all injury and illness out there to the task force members and I never had to clinically see or correct a problem or treat someone, I would consider it a success; in fact, I would consider it our greatest success. That said, there's usually not an environment we get into that allows that.

The fortunate thing about knowing your teammates is that you have a baseline from which to work. Hopefully you will be a step ahead in noticing differences in their overall behavior, whether that's their gait or posture changes, or how they breathe. To some degree, I've always said that when you're a task force doctor, you are a combination of two things—a clinician and an advocate. In some cases, I've been called "Mom." The idea is that you really are watching the whole time, which, in this case, is preventive medicine. The mentality is that the best general never has to fight because his planning is so good.

A responder looks for victims in downtown New Orleans.

The other thing is that the job of a task force doctor does not end after a deployment is finished. We get into some environments where we need to make sure we continue to monitor so the members are healthy and ready to come back the next time. The World Trade Center Health Registry is a classic example of this.

In the old days, the town doctor knew everybody, and he had a personal relationship with the members of that community. As a task force, we are that community. For me, that's one of the more gratifying things about being a task force doctor. The truth is, these are your patients, these are your colleagues, and these are your comrades at the same time. Sometimes that's difficult, but in most ways, it's very gratifying.

Scaling up to the second story of a housing complex near downtown New Orleans, a water-strike team member attempts to access a deck where two victims were seen nearly unconsciousness while lying on a patio. The victims were rehabilitated by medical team specialists before being transported to safety.

▶ Water-strike team members pause for a group photo after rescuing hundreds of people.

Highway ramps are often used as boat docks during flooding events, such as this instance in New Orleans.

Responder Interview

Responder: Jeff Saunders
Position: Operations chief
Search and rescue career: 30 years

Q: In your own words, explain the command function within the incident command system.
A: The total effort for search and rescue is to do the most amount of good for the most amount of people in the least amount of time. As such, it's the command function to be able to put the personnel and resources in the positions that will accomplish that goal. It's knowing the abilities of the personnel that are responding and matching their abilities with the specific tasks, and it's making sure you have the right equipment staged prior to an event. It's also making sure you have excellent planning and are able to adapt and overcome the challenges based on that plan. It's an intimate knowledge of your personnel, of your equipment, and of what you are supposed to do. These management principles pertain to all the various levels of management, whether on the overall task force level or the specific operations level or the area command level.

Q: Hypothetically, a hurricane is going to make landfall on the coast. What is your course of action from start to finish?
A: The first thing we do is determine how many assets we're going to need—things like personnel and equipment, task forces, boats, helicopters, airplanes—whatever it is based on the size and impact area of the storm. Then we stage our resources. We place a small contingency of resources inside a safe area adjacent to the impact zone, so they'll be the first people to pop up and start a rapid-needs assessment. Some of them are going to be staged in the path, and we go in knowing that. We do that because we want them to begin working immediately without travel time. Then, we bring in the rest of the resources to the impacted area based on initial damage assessments and requests from local jurisdictions.

After the storm passes, we're going to be doing two things immediately. We do our own assessment of the situation so we can send information back to the Texas Division of Emergency Management. This includes reconnaissance and rapid-needs assessment specifically for search and rescue. We want to know what the damage looks like and if there are critical needs as far as search and rescue are concerned. Then, we place liaison officers inside emergency operations centers. We get pertinent information of jurisdictional needs from the centers, and we initiate search and rescue activities based on those needs.

After operations are complete, the local jurisdictions notify the Texas State Operations Center that our services are no longer needed, and we are then demobilized.

Q: What are your management and leadership philosophies?
A: My primary philosophy is that ego means "Everyone Gets an Opinion," but at some point the commander needs to make command decisions based on that input. Also, when everything is starting to speed up, I don't believe you can speed up with it. You have to slow down and be very deliberate with decisions. I want to always know what I'm doing and why I'm doing it.

Q: What is your course of action when dealing with local jurisdictions as you come in as the "state"?
A: In the initial meeting we have with a local jurisdiction during a disaster, the two things that stand out in my head are the fact that you're there not only to help but to educate that jurisdiction on what your capabilities and capacities are while you're there. That allows them to understand what you have to offer so they can best fit you within their needs. We tell them what's available, and, as local residents, they can best assess and ask for assistance with what they need. We are very clear on what our capabilities are and are not. We are specifically a search and rescue organization. We don't specialize in a lot of things outside that field, but we may have the resources to assist with something outside that direct field if mandatory to the challenge. Everything outside direct search and rescue is peripheral to our primary mission. It's not that we don't ever work outside that primary mission, but it's not our focus.

Q: What is one of your most vivid memories with the task force?

A responder prepares for UH-60 Blackhawk helicopter water-rescue operations.

A: I remember going into New Orleans directly after Hurricane Katrina and seeing that we had a boat ramp that was a three-lane freeway dropping into the new Gulf of Mexico. It was pretty eye opening. That was the point when we realized this incident was as big as the assets we brought to bear and that all were going to have to step up their game to make it work. Texas Task Force 1 served as both a state and federal asset during Katrina. We sent a team at the request of the Texas governor, and we sent a federal team at the request of FEMA. Our FEMA-requested team was operational inside the city of New Orleans within twenty hours of the hurricane making landfall. Our task force and task forces from Missouri and Tennessee were the first three FEMA-requested search and rescue teams in the city. We were making rescues directly after the flooding with limited communication abilities and completely pushing the envelope of what our normal operations would be in order to get people out of harm's way. We were working nonstop pulling people off houses and out of houses and everywhere else. I still have the brown paper bag that was the first situation report I received because our radio equipment would not transmit far enough in the field at the time. That report said, "Hands poking through roofs; limited food and water; limited communication; actively performing rescues."

5 : Canines

Despite the technological arsenal responders have at their disposal—search devices, cameras, communications equipment—there are still situations when a search cannot be completed as quickly and easily as with a highly trained search and rescue dog. These four-legged responders can search a twenty thousand–foot-high concrete and wood rubble pile in thirty minutes, finding a number of victims and barking upon reaching each one to let their handlers know that person's exact location. Few things are as truly remarkable as watching a search dog move over a sea of unstable wood, nails, and other rubbish while working tirelessly to locate trapped victims.

A great search dog meets a number of set requirements, including those related to disposition, physical agility, size, energy level, and general breed (although there are exceptions). Finding a dog with these requirements is only half the battle, however, because it takes a disciplined and qualified handler to mold the dog into a life-saving responder. Unique to the canine team, both the dog and the handler must be trained together, essentially doubling the commitment of time and training. For federal certification, a series of rigorous and exceedingly difficult tests must be passed, tests that are held only a few times annually.

The Texas Task Force 1 canine program began in 1997, shortly after the team's inception. The founders knew they needed twelve rescue dogs to meet the federal search team requirements, and in these early years they recruited handlers with wilderness search backgrounds and other similar specialties to build the unit's foundation. Susann Brown, a canine specialist who already had several years of experience, was hired to serve as the primary canine program manager. She began the program by developing the initial guidelines and training procedures and by forging the individual handlers into a cohesive response team. As time went on, both Susann and the canine handlers took a fast track to the top in major disaster response, including deployment to the 2001 World Trade Center after the attacks and a massive wide-area search for remnants of the Space Shuttle Columbia after its failed return to Earth in 2003. Texas Task Force 1's canine search program and the team in its entirety quickly became well known for their search and rescue expertise. Canine responses have continued throughout the years with activations to many natural disasters, including Hurricanes Ivan, Katrina, Rita, and Ike.

What Does It Take to Become a Search and Rescue Dog?

The first thing is breed, although specific breeds are more a guideline than a necessary requirement for search dogs. There are no hard-and-fast rules as to what breeds make the best search and rescue canines, but it is generally accepted that certain breeds regularly produce dogs that are better suited for a life of working rubble. In fact, FEMA produces annual statistics of what breeds successfully make it through its extensive testing regimens and become certified rescue dogs. Those breeds often include Labradors, shepherds, retrievers, and other hunting breeds, although there are various other breeds that continuously work well in the field. For example, during the World Trade Center response after 9/11, even pit bulls and rat terriers were used effectively.

The next thing is genetic makeup. Most of us realize that no matter how hard we work out, we may not have the genes to be an Olympic weightlifter. Dogs are the same. Canines in the search and rescue field are asked to perform a very difficult set of skills, one that few dogs are capable of performing. No matter how much the handler or the dog may want to perform, some dogs just physically and/or mentally cannot do it. So, specific programs have been developed to quantify what actually makes a great canine search specialist. Dogs are screened long before they are given their first training exercise. Just like human responders, if the dog passes its

Task force canine handlers perform training exercises near a derailed passenger train while working at Disaster City.

A canine team is ready for anything.

Training is a continuing ritual for handlers and their canines, such as this Labrador (Maggie), who is performing a trench search exercise.

A search and rescue canine sleeps while riding out to the disaster.

It is bath time after a hard day's work.

Search and rescue dogs love to have fun as they work. Although people-friendly, they work for their special toys as they perform searches. They are rewarded with their favorite toy or treat for each victim they find.

Search and rescue canine Kinsey walks with her handler, Bob Deeds, after returning from a response to Hurricane Ivan.

Canine handlers often have very close relationships with their canine search partners. Here, Denise Corliss shares some ice cream with her dog, Bretagne, during a deployment debriefing meeting.

Team members are briefed on the day's response plans.

initial screening tests and job interviews, then it gets to move on to training.

The canine search and rescue training program includes monthly checklists of what a dog should be able to accomplish throughout the regimen. Under the current federal system, a field skills assessment is given to handlers who think their dog is ready for testing. This assessment includes the areas of obedience, directionals, agility, and search. During the search assessment, the dogs must be able to search two rubble piles, each twenty thousand square feet,

Although search and rescue dogs can be relaxed off duty, they are all business when disaster strikes.

and locate numerous victims. Once a dog has made it to this level and passed a number of other assessments, it moves on to advanced screening and testing. To sum up advanced testing in one sentence: Find six victims in three massive rubble piles in less than an hour with limited visual commands from the handler. Think your dog is up to it?

What Does It Take to Become a Search-and-Rescue Canine Handler?

Urban search and rescue is considered the PhD of canine handler training. Most people who enter the field of urban search and rescue canine handling have had prior experience working dogs, whether in wilderness search and rescue, human remains location, narcotics detection, or explosive detection. There are exceptions if a person has the right support and the right dog and everything comes together seamlessly, but this is extremely rare. Most inexperienced handlers have not developed the problem-solving skills usually needed when deep in training.

A major part of a handler's education comes from screening dogs. Handlers learn what they do and do not need in a dog; they learn that if they do not have the right dog, then they are not going to be successful. It takes having both the right dog and the right human skills. Handlers cannot train these dogs by themselves—it is a team effort. They have to spend time on rubble and at training facilities like Disaster City. They have to be disciplined, working with their dogs on skills such as obedience and directionals on an everyday basis.

The canine handler is often considered to be more important in creating a successful search dog than the dog itself. Before new handlers are selected for Texas Task Force 1, they are extensively interviewed on a number of levels. They are assessed not only on their canine-handling abilities but also on their general personality, ethics, and demeanor. During a disaster, canine search handlers may not always be needed during the extent of the disaster response. Therefore, the type of individual needed is one who is going to be setting up tents and moving boxes when not in the field searching. The canine is the tool, but the handler is the craftsman. If the canine tool is not needed at the time, then it is important to be doing what you can with the other tools available.

Texas Task Force 1 chooses canine handlers based on the following criteria: (1) a team player who works selflessly with others, (2) an experienced canine handler who is educated in the field, and (3) someone who has a quality dog (with exceptions). If a potential handler walks in the door and possesses strong qualifications in areas 1 and 2, then he or she may have a chance of being added to the roster and receiving assistance in finding the perfect dog.

Jim Yeager, Texas Task Force 1 search manager, explains, "The first thing I ask people who are interested in becoming a canine handler is if they have any experience. Do you have any search and rescue experience? If not, and you'd like to be on a federal team one day, then get as much experience as you can. Join your local search and rescue team. Learn as much as you can. Read as much as you can. If you want to be a canine handler, I think there are two routes. One path is to get your experience.

It can take years, but you have to be patient. If you haven't been a canine handler before, then begin with a local team and get that experience, develop your skills, and build your resume. Then, you can apply when you're ready. The second path is to join a task force in a position that may not require as much background, such as a logistics specialist. Once on a team, you can learn directly from the canine handlers and build your foundation that way. Try to work with the canines and the handlers as much as you can; learn all you can. Then, one day, when you have the experience and the right dog, it may be a possibility to move into an urban search and rescue canine program."

Canine handlers must learn how to care for their search partner in case of injury in the field, such as during this training exercise.

Canines 93

A handler and her dog show some love after a long day of searching

A handler and her dog work on training while awaiting the landfall of Hurricane Ike.

Texas Task Force 1 canine handlers pose with their dogs after receiving advanced search certification from the Federal Emergency Management Agency.

Responder Interview

Responder: Susann Brown
Position: Search team manager
Search and rescue career: 25 years
Canine partner: Rose

Q: How did you get started in search and rescue as a canine handler?
A: Well, I initially became involved after a personal incident while vacationing at the beach about twenty-five years ago. One of my kids wandered off with my niece, and they got lost. We called the police and beach patrol, and it took about two hours to figure out where they were. It turned out they had just walked farther than expected and couldn't make it back. But those four hours were the most frightening and frustrating of my life. I had no idea what to do. I felt so completely helpless to do anything other than go sit by the phone and wait. I never wanted to be in that situation again. About a year later I was looking to get a new dog, and I was thinking about getting a Labrador because Labradors are generally good with kids, and I like the breed. I also decided I wanted a dog that could do search and rescue because I knew dogs were starting to do that and it peaked my interest. So, after I bought my dog, I found a volunteer search group and got hooked up with them and some other teams throughout the following years. Several years later, I was recruited to join Texas Task Force 1 with the goal of directing the canine program. I actually started the position in 2000 and became the canine program manager. It's been quite a journey since. I've been to 9/11, the Space Shuttle Columbia disaster, Hurricane Katrina—countless disasters where I've been able to use my skills.

Q: So you have responded to the largest terrorist attack and largest natural disaster in our country's history?
A: Yes. You often realize at the time that you're in an event that is of historic proportions. The thing is, you'll always carry stories that no one else will know. You will probably have more of a true reckoning of what occurred than anything you see on television. It seems to me there's a responsibility that goes along with that—a privilege to serve. There's a responsibility as a bearer of those stories to tell them, even if they weren't the main-line stories; they are all important historically.

Q: You have had four dogs. Do they realize they are on deployment searching for people in life-and-death situations?
A: They're playing. They're pretty simple creatures, and they do what they do because they want to do it. We create a world for them that says that anyone they are looking for will have a toy or a treat for them, and that's why they do it. Certainly some of them really like people, but they don't have to like people—they just have to like their toy. They really do it for that reward—that bonus at the end.

Q: When you send your dog into that rubble pile or collapsed structure, it has to pull at your heartstrings. Maybe it is similar to sending your child into a dangerous situation and hoping the child will save the lives of others because the child is agile and small. Do you work on overcoming that mentally during your training?
A: Well, my first canine partner actually died from contamination poisoning on a search. So, I've had that experience of losing a working dog in a line-of-duty

death. It's very real to me—the risk is very real. The potential loss of any dog is there, and it is a devastating loss. I think the hardest part about it is that none of these dogs came in and asked for an application. They don't say, "I want to put my life at risk to do disaster search." We choose them; they don't choose us. It's a bargain you make with them. You say, "Look, I am going to give you a great life. You are going to have more fun than you ever dreamed of, and I'm going to take care of you. I am going to put you at risk, but trust me, I'm going to take care of you." It seems a violation of that trust to lose a dog. But the other side of it is that we know they increase the survivability of people. I love dogs tremendously—maybe even more than some people—but I value people life more, and I am willing to put my dog's life at risk to save people.

Q: It is interesting to me that you say you lost a "canine partner" rather than your dog. Is it a different relationship when you have a canine partner rather than a pet?
A: Yes it is. There's an emotional barrier for a couple of reasons. For one, she has to work for me, not be with me. She has to have such a level of respect for me that when I say do it, she needs to do it, oftentimes for her personal safety. So we unfortunately can't have this touchy-feely, huggy relationship where those lines of obedience and respect are blurred. If they are, it will put her at risk—there's no question about it. So I have learned through the dogs I've had, after I lost my first dog, that there has to be that line . . . I keep myself from falling in love with them. I love each one of them, but I have to draw my personal lines and keep myself from *falling* in love with them. My friends and family can spoil her and play with her in ways that I can't do until she retires. Then I can and will. So there is an emotional barrier for me. I can't say that's true for all handlers, but I strongly feel that way because I have lost a dog and I understand the severity of continuing as a handler after that. This job is important to me, and if I lose a dog, I still want to do my job. If I fell too deeply in love with a dog and lost it, then I may not be able to do my job.

Q: It seems like that would be especially hard for a canine handler because you are in fact a canine handler in the first place. You love dogs.
A: It is. And there are a lot of handlers who don't keep that mental distinction, and I don't know which mind-set is necessarily better, if either. I know it's my survival skills, and that's what works for me.

Q: Where do you find search and rescue canine candidates?
A: They come from all over—they can come from other search and rescue dogs or existing training programs. Quite a few of our dogs actually come out of shelter/rescue programs—some of them seem quite incorrigible because they have a working temperament. Some of them have a hard time being a pet when they have no focus. They are good as pets when they have a job because they can work and work and work; then they get to relax. But if they don't have a job, then they're constantly going to people saying, "Give me something to do, or I'm going to come up with something on my own and you're not going to like it." That's often why they were in animal rescue in the first place—because they are so destructive. This is especially true if they're toy obsessed. Whatever they think the toy is, that's what they are going to go after.

Q: How many dogs from shelter/rescue programs are on the team?
A: Four of the twelve dogs on Texas Task Force 1 are from rescue programs. Often dogs that have a really high drive for reward articles are often considered delinquent because they act like pests. They pester their owners, and they don't handle idle time well. Most people don't want that type of dog. A lot of people want dogs that curl up at your feet and lick your hand, and these dogs just don't do that well. Once the dogs have a job, they get their "fix," so they can rest at other times. These dogs are extremely physically active and in excellent physical condition. They are very agile and live in a three-dimensional world. When you go to do something, they may be up on a table or up on a counter because they love jumping and climbing. That's what we want, and shelters are a good place to find that type of behavior.

Q: Who owns the dogs?
A: Many of the dogs are owned directly by their handler, but some of them are owned by the state.

Q: And how does the state choose a search and rescue canine?
A: Before we buy a dog, I take it home and spend some time with it to make sure we're spending our money wisely and that the dog does what we expect before we make a purchase. I like taking a dog that has potential and pulling up all the conceptualized badness and saying, "Look, you can do all of this stuff as long as you follow these rules. You can be as bad as you want, but you have to direct your energy this way." The dogs who really want to do this job, who really like this job are thinking, "Wow, I've just died and gone to heaven."

Q: What does a search and rescue canine do on a nonworking day as opposed to a training day or when it is on deployment?
A: On a nonworking day, my dog, Rose, comes to work with me and stays by my desk. If I'm out of town on travel, I have people who keep her for me. When we get home from work, I typically exercise her daily by throwing toys. For us, that means we are working on remote-control obedience where I'll send her to different places on the property and give her a toy for doing that. It gives her the exercise she needs and works on those little things that keep her focused and doing what I say, when I say it. After that, she hangs around the house at night. She can be very relaxed with other people and be a goofball, but not with me. I never let her just lie all over me or give her gratuitous petting. I give her more affection than it sounds like, but it's always when I say; she always has to know that it's because I said it was okay. During training days, we get up early, I get her in the car kennel, and she begins rotationals along with the other dogs and solves search problems in buildings or on rubble and works on her agility and directionals. Because she's certified with an additional handler, she also trains with him. On deployment, she's like everyone else. When she gets to the rostering location, she is checked in and gone over by a vet before deploying. When she returns, her health is reviewed again.

Q: What type of diet does Rose have? Does it change on deployment?
A: All the handlers are a little different, but there's a national search-dog foundation that has a program I use that supplies a year's ration of dry food products to federally certified dogs. I vacuum-seal single-serving

sizes of the food along with whatever vitamins I add into it. Those travel with me, and I have twenty days' worth of that plus a ten-pound bag of food. That way, we don't change food if something happens and we're on deployment longer than expected. We could never bring a standard brand for twelve dogs; it would never work. Everyone has a different feeding regimen for his or her dog. I vacuum-seal the bags myself and date them, so on deployment I just have these little bags of food that I can cut open, dump in the bowl, and have that day's supply. If she's using a lot of calories one day, I have extra that I can augment that with. I feed her at the same time every day, so I also relatively know her bowel movements, which is supremely convenient on deployments. She eats about two and one-half cups in the summer and then goes up a half cup in the winter.

Q: How do the canines work in situations like Hurricane Katrina, where everything is underwater. Are there huge contamination issues with them since they cannot really wear protective gear? Do they try to drink the water in flooding events?
A: Katrina was really unique for us in that we had canines on the team for the nearly month timeline we were there, and we didn't deploy our canines in the field once. We were working from boats the first two weeks, and the water was so contaminated that there was no reason to put the dogs in that position. Once they get the contaminants on their bodies, they are going to lick it off and ingest it. The risk was so much higher than the potential benefit. We were personally checking every house, so they weren't going to be of tremendous value during that situation. After the water receded, we were then in these extremely muddy, mucky areas where those contaminants were even more concentrated. So we didn't use them there either. It's always a call between the risk to them and the reward. Certainly, we put them at risk to do their job, but if the benefit doesn't outweigh the risk, then there's no point doing it.

Q: What medical supplies do you keep in your pack for the dogs when directly in the field?
A: We carry a mild dog shampoo and basic first-aid supplies and bandaging material. We all carry fabric muzzles so that if the dogs get injured, we can muzzle them to take care of them when we're handling them. When they are severely injured, they could easily be dangerous even though they don't intend to be. We also carry some basic medications that cross between people and dogs.

Susann Brown works on directional training with her dog, Rose, on a wood and concrete rubble pile at Disaster City.

6 : Disasters

Texas Task Force 1 is unique in that the team has responded to such a wide variety of disaster events throughout the United States. We as a nation experienced the largest natural and human-caused catastrophes in the course of 2001–2005: the September 11, 2001, terrorist attacks on the World Trade Center and Hurricane Katrina. Although they were completely different types of disasters, Texas Task Force 1 responded to both, as well as to others both large and small. From floods to terrorist attacks to hurricanes to earthquakes, Texas Task Force 1 is prepared to respond and will continue its mission to provide safety to the citizens of Texas and the nation.

A Texas Task Force 1 squad assists an elderly man from his New Orleans home after Hurricane Katrina.

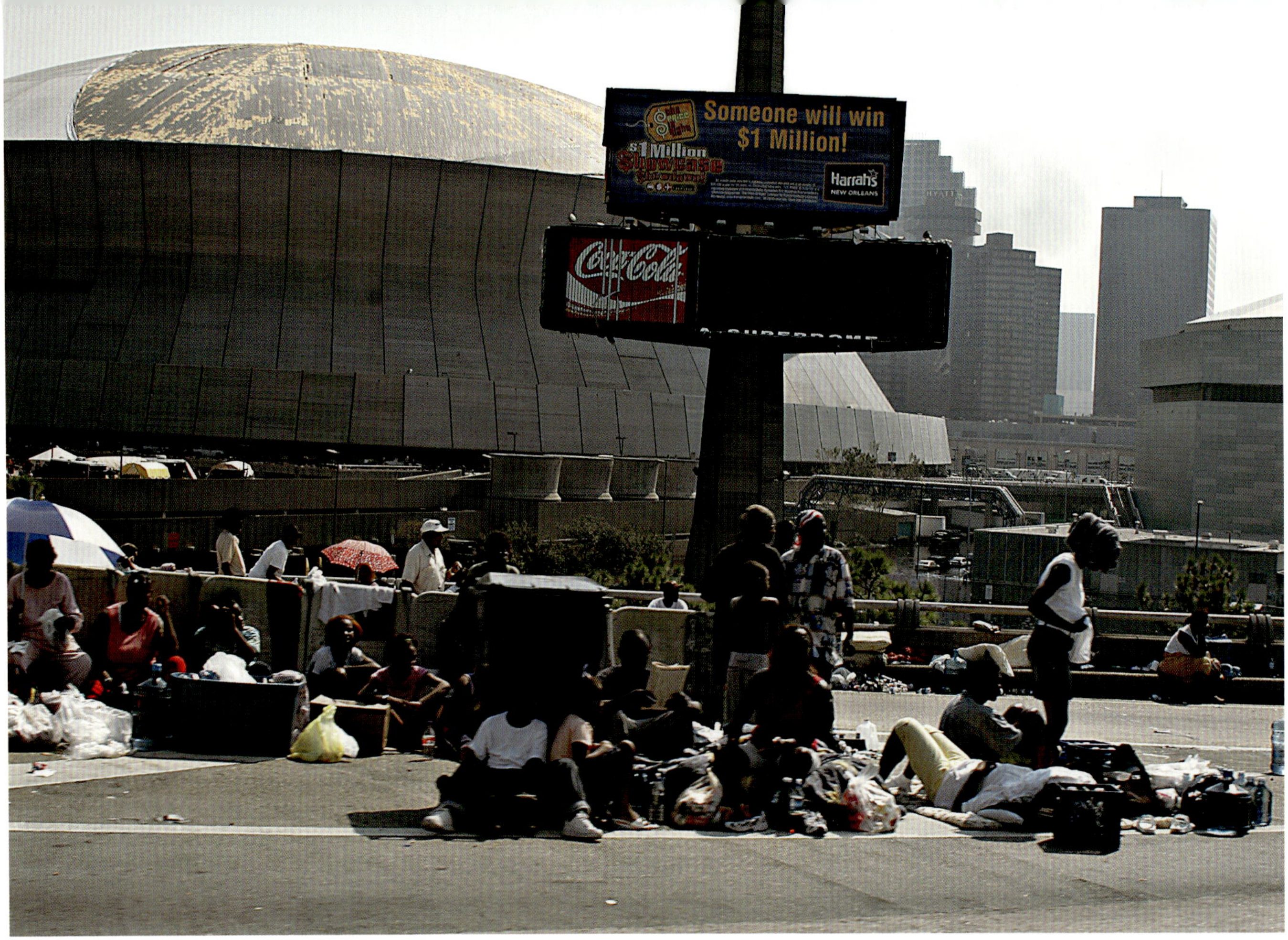
Someone will win
$1 Million!
$1 Million
Showcase
Harrah's
NEW ORLEANS
Coca-Cola

AMTRAK
AMTRAK

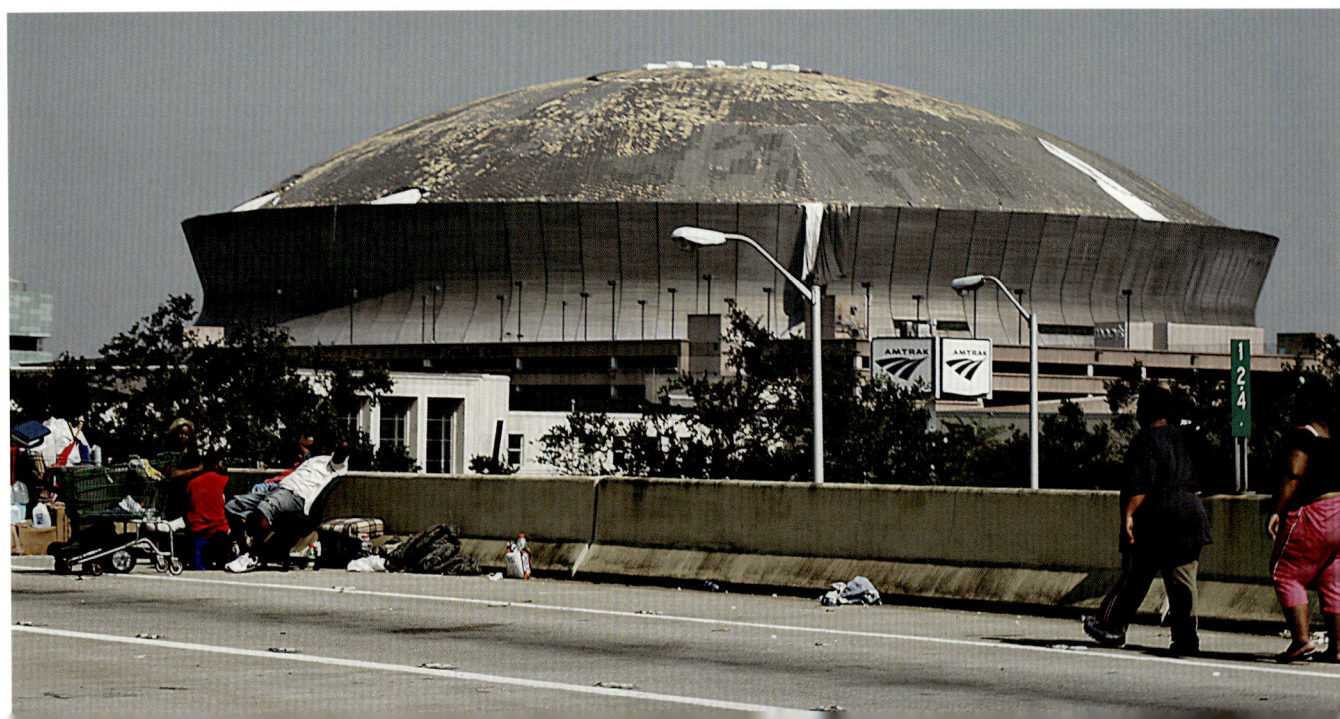

The Super Dome in New Orleans.

A bedsheet from a hotel window signals for help.

Grateful to be rescued.

Overpass evacuation.

Emotions run high as a woman thanks the responders who are assisting her to safety.

Interagency coordination.

Wading through the neighborhoods of New Orleans.

Communicating with other squads to determine which sectors of the city had been covered.

Rescue.

Flooding events such as the devastation in New Orleans often turn normal sights into the surreal.

Responders check a bank's entrance.

A dog awaits rescue from a flooded home.

Both people and animals were seen for days on rooftops around the city following Hurricane Katrina.

Children await dry land while being transported via boat in New Orleans.

Wading through the streets of New Orleans.

Working in the theater of disaster.

Evacuees were transported via water, land, and air during escape from the devastation of Hurricane Katrina.

Evacuees and responders abound in downtown New Orleans.

Texas Task Force 1 and the Texas Forest Service work inside a mobile emergency operations center.

Approaching South Texas to face Hurricane Alex's 2010 landfall.

116 CHAPTER 6

Evacuating residents from Bolivar Peninsula near Galveston Island after Hurricane Ike.

A battered American flag flies over a member of Texas Task Force 1 on Bolivar Penninsula just after he rescued two elderly women from a nearby home.

Awaiting the arrival of a Chinook helicopter to assist in transporting evacuees from Bolivar Peninsula.

UH-60 Blackhawk helicopters are sometimes used to transport the team to remote areas to provide wide-area search and rescue operations.

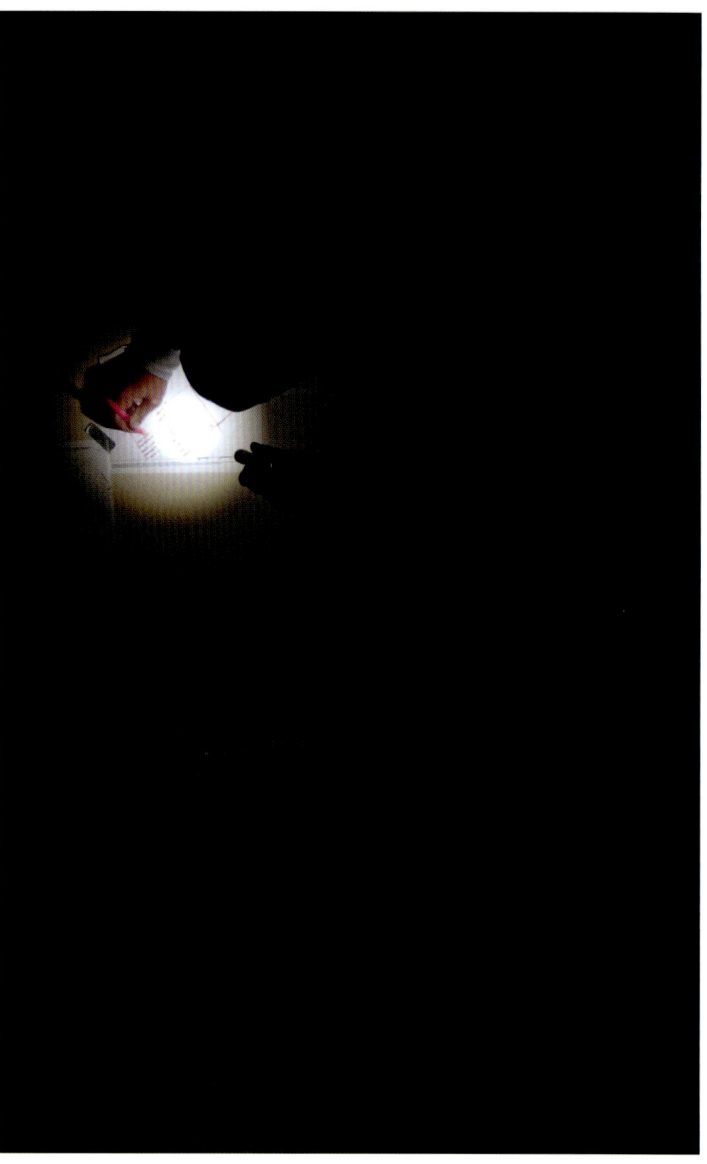

Analyzing search maps by flashlight in the earliest hours of the morning.

▶ Searching rubble for possible victims inside demolished coastal structures.

Crystal Beach one day after Hurricane Ike.

Deciding the quickest possible course of action to search the remaining Bolivar Peninsula after Hurricane Ike.

Evacuating medical special needs patients onto a C-130 aircraft before the landfall of Hurricane Gustav.

A beautiful Galveston Island sunset reflects on its flooded streets.

Task force leader Chuck Jones gives search instructions before leaving for Bolivar Peninsula.

Rescue operations in the waters surrounding Pelican Island.

Asphalt is no match for major storm surges and hurricane-force winds.

▶ Preparing to leave headquarters for Hurricane Alex before daybreak.

Hurricane winds may demolish one structure while leaving another intact only feet away.

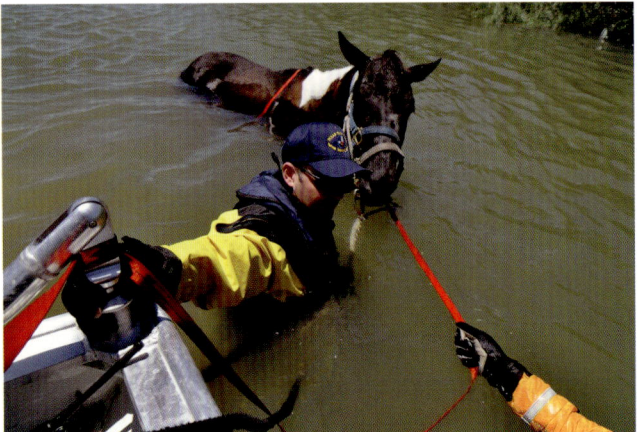

Rescuing a horse from the Rio Grande River after major flooding along the Texas/Mexico border. Although Texas Task Force 1's primary mission is saving people, the team can adapt to other tasks and provide the assistance needed by local jurisdictions.

Falcon Dam releasing water after historic flooding along the Texas border in 2010.

A team reviews search grids within Texas' Cameron County.

Aerial reconnaissance via helicopter.

Disasters

An international border crossing between Texas and Mexico after weeks of hurricane and tropical storm flooding.

Evacuees navigating the flooded Rio Grande.

Disasters 131

The normal path of the Rio Grande River is barely discernible after major flooding.

Rooftops and debris replace houses and lawns along the Rio Grande.

Double- and triple-decker trailers are often used for transporting multiple boats to a single location.

▶ Getting dressed in completely waterproof dry suits for the day's operations.

A wide river replaces the street below.

Analyzing the Rio Grande along the international border.

Street stop signs are not very meaningful when you are in a boat.

Victims are transported to safety on Galveston Island.

Responders access a flooded street at dusk to rescue an injured child after Hurricane Ike ripped through Galveston.

No response is ever made without comprehensive planning.

A team evacuates medical special needs patients onto a C-130 aircraft.

Photographs on this and the following page courtesy Texas Engineering Extension Service (TEEX).

A squad of Texas Task Force 1 responders search unstable rubble for victims in the aftermath of the World Trade Center attacks.

Family members and friends posted mementos, cards, and flags throughout the headquarters of Texas Task Force 1 while members were on deployment to New York City.

A portion of the western wall of Tower 2 stands erect.

Working twenty-four hours a day in various shifts, the task force and other agencies never stopped in their search for victims and recoveries.

An American flag flies over the steel shell that was once the World Trade Center.

Disasters 137

Warren "Country" Weidler and Billy Parker work at Texas Task Force 1's base of operations at Ground Zero.

Gathering a plan of action while searching Ground Zero, squad members worked in small groups and covered specified search area quadrants.

Responders receive a briefing as they head out on their shift's search of Ground Zero.

Due to high levels of airborne and settled surface dust, all Texas Task Force 1 responders wore respirators.

Responders look like specks in the rubble as the early-morning sun shines through the dust and debris, exposing the faint silhouette of the World Trade Center's outer shell.

Many of the Texas Task Force 1 members who were present for the search at the World Trade Center have been deployed to several of the team's other major responses, including the Space Shuttle Columbia search and Hurricane Katrina.

Texas Task Force 1 worked hand in hand with various agencies and organizations to complete an exhaustive search of Ground Zero.

Afterword

J. Robert McKee, director, Texas Task Force 1

As an organization, Texas Task Force 1 has been fortunate to have visionary advisers, former leaders, and incredibly dedicated men, women, and canines. Each one who has passed through the hangars of the task force has left an indelible mark on the organization and contributed to the reputation Texas Task Force 1 enjoys today. While the future continues to evolve for search and rescue, the organization will morph with those changes as it has in the past.

One of the task force's many strengths is the creativity and ingenuity of the members. Throughout each deployment, new challenges have forced task force leaders, managers, and members to think and act uniquely. This has become a hallmark of Texas Task Force 1, as it has been asked to continually assist in additional ways and do so by improvising and adapting to the situation and environment. This forward-thinking approach to emergency response begins at the top with the current governor of Texas, Rick Perry, and his response team. Texas Task Force 1 has tried to answer the call by creating new programs and versatile deployment models and by assisting other jurisdictions in developing search and rescue capabilities.

Countless studies have been written on the challenges faced during Hurricane Katrina in 2005. As a result of this event, major changes in the way emergency response is conducted have ensued. The political pressure faced by elected leaders to respond efficiently and effectively has raised the stakes in a post-Katrina world. The state of Texas takes a proactive and ever-ready stance in its approach to emergencies. TX-TF1 is often activated early in an incident or even before a situation may actually develop. This approach in leaning forward has allowed responders to be positioned early and in the correct staging posture. This can be done only with the support of committed leaders as we have in Texas.

The Texas Task Force 1 membership is awe inspiring indeed, and the commitment and dedication of these brave and selfless responders are unique. A great amount of time and energy is required to become a member; training never ceases and becomes even more important as the months and years pass. At any given moment, one can find the members of Texas Task Force 1 training to improve their skills in finding and rescuing others.

When the task force is not responding, it is training on principles that continue into an actual response. Imagine several hundred captive search and rescue professionals waiting for a hurricane to make landfall. They are ready to use their skills to help other people, but they must wait for the disaster. So, while preparing to respond, they train to remain proficient and able to perform at the highest level possible once the disaster does strike. This is what sets these men and women apart—their desire to help others in the very worst of times.

Behind every task force responder is an incredible support network: first of which is his or her employer or agency. Without that support, TX-TF1 would not be what it is today. Employers with a task force partnership commit the resources in people and help citizens all over the state and nation. The most important network emotionally, however, is family. Texas Task Force 1 could not respond without the family members supporting its responders. The true heroes of the task force are the family members who care for those who save lives. Much as is the case with a military family, sacrifices are made by all in order to conduct the missions to which we are called. One often thinks of the children at home while their mother or father is at Ground Zero on September 11, 2001, or the images of New Orleans in the aftermath of Hurricane Katrina, only to realize their family member is at the heart of that particular event. The family members allow the task force to be what it is today: an extremely important and well-respected response asset.

The future of Texas Task Force 1 and emergency response will undoubtedly be sound. As disasters surface, hurricanes develop, and emergencies continue, the task force will do its part to assist. How that mission is accomplished will change, however, as emerging technologies enter into the plethora of resources that search and rescue utilize. From robotics, to image sensors, to communication resources, our responders continue to have new technological advances with each deployment. The ability to put that technology to work will not only save the lives of the victims involved in a disaster but will save lives of the responders as well. Known hazardous conditions, for instance, can be mitigated through the use of remotely operated vehicles as opposed to sending a human being into a life-threatening environment. The future is an opportunity to utilize new tools to further help rescue citizens in harm's way and to protect the environment where appropriate.

As one reflects, this book has provided the reader with a unique perspective on the first twelve years of TX-TF1. The next twelve years will be full of change and opportunity. Members will come and go, but the important mission of this organization will not change: help where possible and provide state-of-the-art solutions to chaotic and dangerous situations. Through the skill and tenacity of its members, TX-TF1 has a bright and promising future ahead.

The journey continues.

Index

Aggie bonfire collapse, 7
Alex. *See* hurricanes
Alfred P. Murrah building, ix, 3, 8
automated response program, 52

base of operations (BOO), 19, 34, 38, 40, 51, 52, 54, 68
Bennett, G. Kemble, viii, ix, x, 3
Blue Bell Ice Cream, 51
Bolivar Peninsula, 38, 65, 117, 119, 123
Brayton Fire Training Field, 4, 12
breaching and breaking, 12, 15, 17, 19, 30, 53
Bretagne, 89
Brown, Susann, 68, 82, 95–99

Cache. *See* equipment
camaraderie, 69
Cameron County, 129
canine: 60, 82, 84, 86–87, 88, 91, 93, 109; handlers, 1–2, 52, 82, 84, 86, 89, 92, 96; candidates, 96; search teams, 5, 8, 85, 89; federal certification, 82, 90, 92, 94; screening, 92; program, 82, 90; breeds, 84; toys, 88, 95; program manager, 95; diet, 97, 98; training, 99
capabilities, 8

chemical, biological, radiological, nuclear, and explosion (CBRNE) 7
Coast Guard, 73
Cockrell, Michael, 59, 65
Comal River, 6
command, 5, 8
command function, 79
confined space, 8, 71
contamination, 41, 98
contaminated environment, 5, 57, 138
convoy, 52, 58
Corliss, Denise, 89
Crystal Beach, 38, 63, 122

decontamination, 41
Deeds, Bob, 66, 88
Del Rio, Texas, 7
Department of Homeland Security (DHS), ix, 1, 23
DHS. *See* Department of Homeland Security
Disaster City, 9, 12, 14–17, 19–20, 28, 29, 84, 92, 99
dry suits, 40, 132

earthquakes, 8, 10–11, 52, 65, 100
East Texas, 66

Emergency Operations Center (EOC), 80, 115
EOC. *See* Emergency Operations Center
equipment, 4, 6, 8–9, 14–16, 31, 36, 38, 40, 43–48, 52, 82
explosive detection, 92
extricate, 23, 37
evacuees, 113–114

Falcon Dam, 129
federal asset, 81
federal: teams, 3, 42, 50; agencies, 6; system, 50
Federal Emergency Management Agency (FEMA), ix, 1, 8, 81, 84, 94
FEMA. *See* Federal Emergency Management Agency
firefighters, 1–2
firefighting school, 4
forest service, 49
full scale exercise, 10, 19, 22, 26

Galveston Island, 38, 63–64, 117–118, 124, 134
global positioning system, 34
GPS. *See* global positioning system
Ground Zero, 51, 71, 72

Gulf Coast, 34, 81, 123
Gustav. *See* hurricanes

hazardous materials (HAZMAT), 3, 5, 14, 41, 46, 71
HAZMAT. *See* hazardous materials
high angle rescue, 19, 25, 71
horse rescue, 128
Houston, Texas, 7, 32, 52
human remains location, 92
hurricanes: 8, 25, 65, 140; Ike: 2, 4, 7, 56, 63–65, 68, 80, 82, 94, 100, 117, 122, 123; Alex: 34, 116, 126, 134; Ivan: 7, 82, 88; Katrina: x, 2, 7, 37, 40, 51, 65–66, 68, 70, 72, 81–82, 95, 98, 100, 110, 113, 139, 140; Rita: 7, 82; Gustav: 2, 32

ICS. *See* Incident Command System
Ike. *See* hurricanes
Incident Command System (ICS), 79
inception, ix, 3, 49
incident command, 31
Ivan, *See* hurricanes

Jones, Chuck, 49–51, 69, 125

Katrina. *See* hurricanes
Kinsey, 88

Labrador, 84, 86, 95
Larsen, Ken, 61, 65
large scale search and rescue team, 5

liaison officers, 80
lifting and moving, 12, 14
logistics, 8, 19, 33, 37, 49, 74, 93

mass casualty disasters, 8
Mathison, Eddie, 76
McKee, J. Robert, 139–140
medical: screening, 4; doctors, 1–3; specialist, 2, 24; team, 8; capabilities, 5; kit/supplies, 74, 98
medicine, 70
Mexico earthquake, 10–11
Migala, Alexandre, 56
military, 2, 42, 49, 52, 56, 58, 66
military aircraft, 2, 38, 42, 50, 52, 56, 58, 61, 80–81, 119–120, 123, 136
Minson, Matthew, 69–75
motivation, 69
Meal, Ready to Eat, 54
MRE. *See* Meal, Ready to Eat

narcotics detection, 92
National Search Dog Foundation, 97
New Orleans, Louisiana, x, 2, 37, 40, 51, 65, 66, 68, 72–73, 76, 78, 81, 100, 103, 108, 111–112, 114, 140
New York City, viii, 7, 51, 71
National Institute of Standards and Technology (NIST), 23
National Urban Search and Rescue System, 3, 8

NIST. *See* National Institute of Standards and Technology

Oklahoma City bombing, ix, 3, 8, 10, 19
Olympics, 7
operations chief, 79

pancake collapse, 52
Parker, Billy, 69, 70, 138
Pelican Island, 125
Perry, Rick, 139
personal floatation device (PFD), 8
PFD. *See* personal floatation device
pit bull, 84
plans, 5
pneumatic shore, 13, 17, 54
processing, 52, 59

rapid needs assessment, 80
rat terriers, 84
reconnaissance, 51, 31, 33, 40, 80, 129
Reliant Center, 32
respiratory protection, 57, 71
retrievers, 84
Rio Grande, 7, 128, 131–133
Rita. *See* hurricanes
robots, 23, 140
Rose, 95, 97, 99
rubble pile, 25, 28, 54, 82, 90, 95

satellite communication, 41
Saunders, Jeff, 79–81
search team manager, 95

September 11, 2001, viii, 2, 7, 100, 140
shelter/rescue programs, 96–97
shepherds, 84
shoring, 12–13, 16, 17, 52
situation report, 81
South Texas, 116
special needs patients, 2, 56, 136
Space Shuttle Columbia, x, 7, 74, 82, 95
sponsoring agencies, 2, 4
state asset, 81
State Operations Center, Texas, 33, 80
structural collapse: 5, 8, 12, 31; specialist, 52; rescue, 71
structural engineers, 2
Super Dome, 103
support network, 140
swift water rescue. *See* water

Texas A&M University (System), x, 3–4

task forces: Missouri, 81; Tennessee, 81
Texas Engineering Extension Service (TEEX), ix, 3–5, 12, 23
TEEX. *See* Texas Engineering Extension Service
Texas Division of Emergency Management (TDEM), 1, 4, 80
TDEM. *See* Texas Division of Emergency Management
Texas Forest Service, 115
Texas Governor, 81, 139
Texas Military Forces (TMF), 6, 38, 59
TMF. *See* Texas Military Forces
tools. *See* equipment
tornadoes, 7–8, 16
trench rescue, 13, 86
triage, 73, 75
Tropical Storm Allison, 7, 71
Technical Skills Training Area, 12, 17

United Kingdom, 23, 27

volunteers, 17, 20–21, 28

water: rescue teams, 5, 8, 26, 64, 76; program, 6, 8, 71; boat squads, 5; training, 19, 26; specialists/technicians, 6, 31, 47; operations, 81
Weidler, Warren (Country), 138
wide area search, 3, 7
wilderness search and rescue, 92
weapons of mass destruction (WMD), 3, 5, 7, 27
WMD: *See* weapons of mass destruction
World Trade Center: viii, x, 2, 51, 71–72, 82, 84, 95, 100, 137, 140; health registry, 76

Yeager, Jim, 68, 92–93